Concise : Inorganic Chemistry

O.P. Thakur

AMIGA PRESS. INC
Delhi -110051

AMIGA PRESS. INC
H.O :- 28-A, Indra Park, Chander Nagar
Delhi :- 110051 (India)
Tel :- 011-22504028, +91-98736-57000
Email :- amigapress@gmail.com

This edition is published by Amiga Press.Inc

ISBN : 978-93-84533-16-8

First Published 2018

Printed and bound in India by Thomson Press Ltd., New Delhi

Concise : Inorganic Chemistry

Preface

Inorganic chemistry is concerned with the properties and reactivity of all chemical elements. Advanced interests focus on understanding the role of metals in biology and the environment, the design and properties of materials for energy and information technology, fundamental studies on the reactivity of main group and transition elements, and nanotechnology. Synthetic efforts are directed at hydrogen storage materials and thermoelectrics, catalysts for solar hydrogen generation, fullerenes and metal porphyrins, metal clusters and compounds with element-element bonds, as well as nanowires and nanoparticles.

Inorganic chemistry, the study of all the elements and their compounds with the exception of carbon and its compounds, which fall under the category of organic chemistry. Inorganic chemistry is significant in all modern areas of life and science. It represents one of the fastest growing and exciting areas of scientific research. The impetus for this research is the practical essential role inorganic chemistry plays in biology, pharmaceuticals, and advanced materials.

Chemistry is the scientific study of interaction of chemical substances that are constituted of atoms or the subatomic particles: protons, electrons and neutrons. Atoms combine to produce molecules or crystals. Chemistry is often called "the central science" because it connects the other natural sciences, such as astronomy, physics, material science, biology, and geology.

In chemistry, a coordination complex or metal complex, is a structure consisting of a central atom or ion (usually metallic), bonded to a surrounding array of molecules or anions (ligands, complexing agents). The atom within a ligand that is directly bonded to the central atom or ion is called the donor atom. Polydentate (multiple bonded) ligands can form a chelate complex. A ligand donates at least one pair of electrons to the central atom/ion. Compounds that contain a coordination complex are called coordination compounds. The central atom or ion, together with all ligands form the coordination sphere.

An atom is composed of three different types of particles; protons, neutrons, and electrons. Because atoms are neutrally charged, an atom will always have the same number of protons and electrons. This idea that atoms are neutral is by definition. Atoms always have the same number of negative and positive charges. By the way, atoms become ions when they become charged. This change can occur when an atom has either lost or gained some electrons. So ions do not have the same number of protons and electrons.

A balanced chemical equation is intended to represent the net transformation of reactants to products that occurs when reaction takes place. It does not, and is not intended to, tell us anything about the details of the reaction pathway. In fact, most reactions occur via a series of relatively simple steps, each involving reaction of only a few (one, two, and very occasionally, three) molecules. Each of these simple steps is called an elementary process. The equation for an elementary process represents an actual collision event at the molecular level. The sequence of elementary processes by which a reaction is proposed to occur is called the mechanism for the reaction.

Molecular self-assembly is the construction of systems without guidance or management from an outside source (other than to provide a suitable environment). The molecules are directed to assemble through noncovalent interactions. Self-assembly may be subdivided into intermolecular self-assembly (to form a supramolecular assembly), and intramolecular self-assembly (or folding as demonstrated by foldamers and polypeptides). Molecular self-assembly also allows the construction of larger structures such as micelles, membranes, vesicles, liquid crystals, and is important to crystal engineering.

The present book is an effort to systematic analyze on the subject.

—Editor

Contents

1

Introduction

Inorganic chemistry is concerned with the properties and reactivity of all chemical elements. Advanced interests focus on understanding the role of metals in biology and the environment, the design and properties of materials for energy and information technology, fundamental studies on the reactivity of main group and transition elements, and nanotechnology. Synthetic efforts are directed at hydrogen storage materials and thermoelectrics, catalysts for solar hydrogen generation, fullerenes and metal porphyrins, metal clusters and compounds with element-element bonds, as well as nanowires and nanoparticles.

Inorganic chemistry is a branch of chemistry that deals with the properties and behavior ofinorganic compounds. Inorganic compounds are generally those that are not biological, and characterized by *not* containing any hydrogen and carbon bonds. It is almost easier to discuss this field in terms of what it is not — organic chemistry. Organic chemistry is the study of any chemical reaction that involves carbon, which is the element that all life is based on.

The term "organic" has traditionally referred only to animal or plant matter, so there is a common misconception that organic chemistry always refers to life processes, or thatinorganic chemistry applies to everything that does not. This assumption is inaccurate. Many chemical processes veer away from this line of thinking, and there are many chemical life processes that depend on inorganic chemical processes.

There are exceptions to every rule. Although carbon is the main common element in organicchemistry, inorganic chemical compounds can contain carbon, too. For example, carbon monoxide and carbon dioxide both contain carbon, but are inorganic compounds. Carbon dioxide, in particular, is also very important to chemical processes necessary for life, especially plant life. The truth is that the lines between inorganic and organic chemistry are often blurred. There are many branches of inorganic chemistry available for specialization. Geochemistry is the study of the chemicals of the Earth and other planets, and it covers the chemical compositions of rocks and soil. Within the field of geochemistry, there are several subfields, including isotope geochemistry, cosmochemistry and biogeochemistry.

Another branch is physical chemistry, which relates to the concept of physics in chemical systems. This field is also sometimes called physicochemistry. It uses the principles ofthermodynamics, quantum chemistry, and kinetics as its basis.

On the other hand, bioinorganic chemistry is the study of compounds containing metal-carbon bonds within biological systems. This is a particularly interesting branch because it also incorporates aspects of organic chemistry into it. Bioinorganic chemistry focuses on the pretense of metal ions in biochemical processes.

Inorganic chemistry lends itself to many different industries, including education, environmental science, and government agencies. A scientist who focuses on this field might create or improve formulas for household cleansers. He may also work in chemical research, coming up with new ways to manipulate the properties of metallic elements into useful functions.

GROUPS AND PERIODS

Inorganic chemistry covers an extremely wide range of chemicals and it is classified according to similar groups, periods and properties. So, for example, the compounds of elements in the same group or period of the periodic table are compared. This leads to the study of the reactions of the Alkali metals (group 1), or the study of the halogens (group 7).

Ions and Electochemistry

One aspect of many inorganic chemicals is the formation of ionic compounds. Ions are charged particles which combine with oppositely charged particles to form compounds. These include chemicals such as Sodium Chloride (table salt) and Ammonium Nitrate (fertilizer). As solids these compounds do not conduct electricity, but as solutions they do and can react when an electric current passes through them. Electrochemistry uses this phenomenon to synthesise certain important chemicals such as aluminium metal and also the principle is used in batteries.

Acid-Base Reactions

Another important area of inorganic chemistry is that of Acids and Bases. Acids are chemicals which have hydrogen ions (H+). Sulphuric Acid is an important industrial chemical which is produced in huge quantities worldwide. Bases accept hydrogen ions readily and lead to further important chemicals.

Transition Metals

The chemistry of the Transition elements, the central block in the periodic table, is vast and important. Many of these elements have large atoms and ions and are prone to form complex ions. A complex occurs when molecules that have weak charges (e.g. water or ammonia) surround a positively charged ion in solution. The resulting complex ion exhibits interesting reactions. This area of inorganic chemistry is called coordination chemistry.

Organometallic and Bioinorganic

The boundary between inorganic and organic chemistry is blurred in the disciplines of organometallic chemistry and bioinorganic chemistry. It is in these areas that important work is being done on treatments of diseases, catalysts and superconductors.

Group 4

One group of particular interest is group four. The elements in this group are carbon, silicon, germanium, tin and lead. In contrast to groups 1, 2 and 7, the properties of the elements in this group change from non-metallic to metallic as the group is descended. These elements, too, are important in the study of superconductors. Silicon is an important chemical, the main constituent of sand and glass and used in applications from breast implants to silly putty!

CHARACTERIZATION OF INORGANIC COMPOUNDS

Because of the diverse range of elements and the correspondingly diverse properties of the resulting derivatives, inorganic chemistry is closely associated with many methods of analysis. Older methods tended to examine bulk properties such as the electrical conductivity of solutions, melting points, solubility, and acidity. With the advent of quantum theory and the corresponding expansion of electronic apparatus, new tools have been introduced to probe the electronic properties of inorganic molecules and solids. Often these measurements provide insights relevant to theoretical models. For example, measurements on the photoelectron spectrum of methane demonstrated that describing the bonding by the two-center, two-electron bonds predicted between the carbon and hydrogen using Valence Bond Theory is not appropriate for describing ionisation processes in a simple way. Such insights led to the popularization of molecular orbital theory as fully delocalised orbitals are a more appropriate simple description of electron removal and electron excitation.

Commonly encountered techniques are:

- X-ray crystallography: This technique allows for the 3D determination of molecular structures.

- Dual polarisation interferometer: This technique measures the conformation and conformational change of molecules.
- Various forms of spectroscopy
- Ultraviolet-visible spectroscopy: Historically, this has been an important tool, since many inorganic compounds are strongly colored
- NMR spectroscopy: Besides ^{1}H and ^{13}C many other "good" NMR nuclei (e.g., ^{11}B, ^{19}F, ^{31}P, and ^{195}Pt) give important information on compound properties and structure. Also the NMR of paramagnetic species can result in important structural information. Proton NMR is also important because the light hydrogen nucleus is not easily detected by X-ray crystallography.
- Infrared spectroscopy: Mostly for absorptions from carbonyl ligands
- Electron nuclear double resonance (ENDOR) spectroscopy
- Mössbauer spectroscopy
- Electron-spin resonance: ESR (or EPR) allows for the measurement of the environment of paramagnetic metal centres.
- Electrochemistry: Cyclic voltammetry and related techniques probe the redox characteristics of compounds.

Synthetic Inorganic Chemistry

Although some inorganic species can be obtained in pure form from nature, most are synthesized in chemical plants and in the laboratory.

Inorganic synthetic methods can be classified roughly according the volatility or solubility of the component reactants. Soluble inorganic compounds are prepared using methods of organic synthesis. For metal-containing compounds that are reactive toward air, Schlenk line and glove box techniques are followed. Volatile compounds and gases are manipulated in "vacuum manifolds" consisting of glass piping interconnected through valves, the entirety of which can be evacuated to 0.001 mm Hg or less. Compounds are condensed using liquid nitrogen (b.p. 78K) or other cryogens. Solids are typically prepared using tube furnaces, the reactants and products being sealed in containers, often made of fused silica (amorphous SiO_2) but sometimes more specialized materials such as welded Ta tubes or Pt "boats". Products and reactants are transported between temperature zones to drive reactions.

THERMODYNAMICS AND INORGANIC CHEMISTRY

An alternative quantitative approach to inorganic chemistry focuses on energies of reactions. This approach is highly traditional and empirical, but it is also useful. Broad concepts that are couched in thermodynamic terms include redox potential, acidity, phase changes. A classic concept in inorganic thermodynamics is the Born-Haber cycle, which is used for assessing the energies of elementary processes such as electron affinity, some of which cannot be observed directly.

Mechanistic Inorganic Chemistry

An important and increasingly popular aspect of inorganic chemistry focuses on reaction pathways. The mechanisms of reactions are discussed differently for different classes of compounds.

Main Group Elements and Lanthanides

The mechanisms of main group compounds of groups 13-18 are usually discussed in the context of organic chemistry (organic compounds are main group compounds, after all). Elements heavier than C, N, O, and F often form compounds with more electrons than predicted by the octet rule, as explained in the article on hypervalent molecules. The mechanisms of their reactions differ from organic compounds for this reason. Elements lighter than carbon (B, Be, Li) as well as Al and Mg often form electron-deficient structures that are electronically akin to carbocations. Such electron-deficient species tend to react via associative pathways. The chemistry of the lanthanides mirrors many aspects of chemistry seen for aluminium.

Transition Metal Complexes

Mechanisms for the reactions of transition metals are discussed differently from main group compounds. The important role of d-orbitals in bonding strongly influences the pathways and rates of ligand substitution and dissociation. These themes are covered in articles on coordination chemistry and ligand. Both associative and dissociative pathways are observed.

An overarching aspect of mechanistic transition metal chemistry is the kinetic lability of the complex illustrated by the exchange of free and bound water in the prototypical complexes $[M(H_2O)_6]^{n+}$:

$$[M(H_2O)_6]^{n+} + 6\ H_2O^* \rightarrow [M(H_2O^*)_6]^{n+} + 6\ H_2O$$

where H_2O^* denotes isotopically enriched water, e.g., $H_2{}^{17}O$

The rates of water exchange varies by 20 orders of magnitude across the periodic table, with lanthanide complexes at one extreme and Ir(III) species being the slowest.

Redox Reactions

Redox reactions are prevalent for the transition elements. Two classes of redox reaction are considered: atom-transfer reactions, such as oxidative addition/reductive elimination, and electron-transfer. A fundamental redox reaction is "self-exchange", which involves the degenerate reaction between an oxidant and a reductant. For example,permanganate and its one-electron reduced relative manganate exchange one electron:

$$[MnO_4]^- + [Mn^*O_4]^{2-} \rightarrow [MnO_4]^{2-} + [Mn^*O_4]^-$$

Reactions at Ligands

Coordinated ligands display reactivity distinct from the free ligands. For example, the acidity of the ammonia ligands in $[Co(NH_3)_6]^{3+}$ is elevated relative to NH_3 itself. Alkenes bound to metal cations are reactive toward nucleophiles whereas alkenes normally are not.

The large and industrially important area of catalysis hinges on the ability of metals to modify the reactivity of organic ligands. Homogeneous catalysis occurs in solution and heterogeneous catalysis occurs when gaseous or dissolved substrates interact with surfaces of solids. Traditionally homogeneous catalysis is considered part of organometallic chemistry and heterogeneous catalysis is discussed in the context of surface science, a subfield of solid state chemistry.

But the basic inorganic chemical principles are the same. Transition metals, almost uniquely, react with small molecules such as CO, H_2, O_2, and C_2H_4. The industrial significance of these feedstocks drives the active area of catalysis. Ligands can also undergo ligand transfer reactions such astransmetalation.

THEORETICAL INORGANIC CHEMISTRY

An alternative perspective on the area of inorganic chemistry begins with the Bohr model of the atom and, using the tools and models oftheoretical chemistry and computational chemistry, expands into bonding in simple and then more complex molecules.

Precise quantum mechanical descriptions for multielectron species, the province of inorganic chemistry, is difficult. This challenge has spawned many semi-quantitative or semi-empirical approaches including molecular orbital theory and ligand field theory, In parallel with these theoretical descriptions, approximate methodologies are employed, including density functional theory.

Exceptions to theories, qualitative and quantitative, are extremely important in the development of the field. For example, $Cu^{II}_2(OAc)_4(H_2O)_2$ is almost diamagnetic below room temperature whereas Crystal Field Theory predicts that the molecule would have two unpaired electrons. The disagreement between qualitative theory (paramagnetic) and observation (diamagnetic) led to the development of models for "magnetic coupling." These improved models led to the development of new magnetic materials and new technologies.

Qualitative Theories

Inorganic chemistry has greatly benefited from qualitative theories. Such theories are easier to learn as they require little background in quantum theory. Within main group compounds, VSEPR theory powerfully predicts, or at least rationalizes, the structures of main group compounds, such as an explanation for why NH_3 is pyramidal whereas ClF_3 is T-shaped. For the transition metals, crystal field theoryallows one to understand the magnetism of many simple complexes, such as why $[Fe^{III}(CN)_6]^{3-}$ has only one unpaired electron, whereas $[Fe^{III}(H_2O)_6]^{3+}$ has five. A particularly powerful qualitative approach to assessing the structure and reactivity begins with classifying molecules according to electron counting, focusing on the numbers of valence electrons, usually at the central atom in a molecule.

Molecular Symmetry Group Theory

A central construct in inorganic chemistry is the theory of molecular symmetry. Mathematical group theory provides the language to describe the shapes of molecules according to their point group symmetry. Group theory also enables factoring and simplification of theoretical calculations.

Spectroscopic features are analyzed and described with respect to the symmetry properties of the, *inter alia,* vibrational or electronic states. Knowledge of the symmetry properties of the ground and excited states allows one to predict the numbers and intensities of absorptions in vibrational and electronic spectra. A classic application of group theory is the prediction of the number of C-O vibrations in substituted metal carbonyl complexes. The most common applications of symmetry to spectroscopy involve vibrational and electronic spectra.

As an instructional tool, group theory highlights commonalities and differences in the bonding of otherwise disparate species, such asWF_6 and $Mo(CO)_6$ or CO_2 and NO_2.

ADVANCED INORGANIC CHEMISTRY

Apart from the above mentioned common topics studied under inorganic chemistry, the new and advanced inorganic chemistry deals with many other topics. Advanced inorganic chemistry is related to many topics from biology, physical chemistry, organic chemistry, etc.

- Chemistry of alloys, clathrates, isopolyacids and heteropolyacids, etc are studied under advanced inorganic chemistry.
- Fuels and furnaces: This topic in inorganic chemistry studies the types of fuel available, both natural and artificial, their chemical and physical

properties, essential requirements of a fuel etc. The use of fuels in furnaces, types of furnaces etc are dealt with under this topic of advanced inorganic chemistry. This topic provides a link to the engineering chemistry and applied chemistry.

- Spectroscopy: This topic is widely learned in physical, organic and inorganic chemistry too. Though its a part of analytical chemistry, spectroscopy and different topics related to it are learned as advanced inorganic chemistry too.
- Organometallic compounds: The term organometallic is usually restricted to those compounds which have a metal atom (including non metallic elements like Boron, silicon, phosphorus, arsenic, etc), which are less electronegative than carbon, and is bonded directly to a carbon atom of at least one organic group radical or molecule. Organometallic compounds are studied under advanced inorganic chemistry and it forms a link to coordination chemistry, chemical bonding, etc.
- Inorganic Polymers: Polymers are usually studied widely under organic chemistry. But there are certain polymers, called 'Inorganic polymers'. All macromolecules which are not having carbon in their backbone are called inorganic polymers. Molecules like silica, graphite, and many other covalently bonded crystals like halides and oxides come under this topic, so study of inorganic polymers is taken care of under advanced inorganic chemistry.
- Study of lanthanides and actinides: The ' f ' block in the periodic table is complicated because of it's incomplete nature. This is usually dealt with under advanced inorganic chemistry.
- Corrosion and passivity: This topic is again a link between inorganic and applied or engineering chemistry, and is dealt with under advanced inorganic chemistry.
- There are many metals, which find their way into the biological systems. They are Sodium, Iron, Magnesium, Potassium etc. The topic inorganic elements in biological systems bridges the gap between inorganic chemistry and biology. It is the latest inclusion under advanced inorganic chemistry. It studies real life complexes like hemoglobin, chlorophyll, etc.
- Thus, inorganic chemistry deals with a lot of topics and is interlinked to biology, other parts of chemistry and engineering.

MEASURE OF REACTION RATES

The rate of reaction is a measure of the change in concentration of reactants or products over time. Rate can be measured at the beginning of the reaction, which is called the *initial rate,* at any point in time while the reaction is in progress, called *instantaneous rate,* or over an interval of time, which is the *average rate*. In the *initial rate* the change in concentration of a reactant or product as a function of time is measured within minutes (or seconds) the reaction starts. Determination of the dependence of *initial rate* to the concentrations of reactants allows one to derive the *rate law* for the reaction. A *rate law* is a mathematical equation that shows the dependence of reaction rate to the molar concentrations of reactants at constant temperature.

The *instantaneous rate* is calculated from the slope of a *tangent* drawn at any points on the *graph of concentrations versus time*. The slope of tangent taken at the initial point of the graph is assumed to be equal to its *initial rate*.

The *average rate* of reaction is obtained by dividing the change in concentration of a reactant or product that occurs over a longer period of time by the time interval that the change occurs. For example, consider the following reaction at 300°C:

$$2\ NO_2(g) \rightarrow 2\ NO(g) + O_2(g)$$

If in the first 150 seconds, the concentration of NO_2 has decreased from 0.0100 mol/L to 0.0055 mol/L, the average rate for the disappearance of NO_2 for first 150 s under this condition is,

$$\frac{-\Delta[NO_2]}{\Delta t} = \frac{-(0.0055\ \text{mol/L} - 0.0100\ \text{mol/L})}{150\ \text{s}} = \frac{0.0045\ \text{mol/L}}{150\ \text{s}} = 3.0 \times 10^{-5}\ \text{mol/(L.s)}$$

Suppose that in the next 150 seconds the concentration of NO_2 decreases from 0.0055 mol/L to 0.0038 mol/L. The average rate during the second 150 s period is,

$$\frac{-\Delta[NO_2]}{\Delta t} = \frac{-(0.0038\ \text{mol/L} - 0.0055\ \text{mol/L})}{150\ \text{s}} = \frac{0.0017\ \text{mol/L}}{150\ \text{s}} = 1.1 \times 10^{-5}\ \text{mol/(L.s)}$$

The *average rate* is the *mean* of several *instantaneous rates* taken over a period of time. Note that *instantaneous rates* of a reaction decreases as time progresses as the concentrations of reactants decrease. Then the average rate taken over a longer period would have a smaller value compared to those taken over a shorter period after the reaction begins.

The rate of reaction may be determined by measuring the rate of disappearance (decrease in concentration) of one of the reactants or the rate of formation (increase in concentration) of a product. For example, in the reaction:

$$2\ NO_2(g) \rightarrow 2\ NO(g) + O_2(g)$$

Various expression of rate can be derived, such as,

$$\text{Rate}_1 = \frac{-\Delta[NO_2]}{\Delta t};\ \text{Rate}_2 = \frac{\Delta[NO]}{\Delta t};\ \text{or Rate}_3 = \frac{\Delta[O_2]}{\Delta t}$$

The stoichiometric relationships between the different rates are as follows:

$$\frac{-\Delta[NO_2]}{\Delta t} = \frac{\Delta[NO]}{\Delta t} = \frac{2\Delta[O_2]}{\Delta t}$$

That is, in this reaction the rate of appearance of NO is the same as the rate of disappearance of NO_2, but double the rate of appearance of O_2.

Rate Laws

Consider a general reaction: $A + B \rightarrow C$

The rate law is expressed as, Rate = $k[A]^x[B]^y$,
where k is the rate constant; x and y are the *rate orders* with respect to individual reactants. The rate orders x and y are not related to the reaction coefficients. Their values must be experimentally determined from a set of kinetic data. For example, the rate law for reaction:

$$2\,N_2O_5(g) \rightarrow 4\,NO(g) + O_2(g), \text{ is Rate} = k[N_2O_5]$$

That is, the rate of the decomposition of N_2O_5 is *first order* with respect to $[N_2O_5]$, which means that the rate will double if $[N_2O_5]$ is doubled; it will triple if $[N_2O_5]$ is tripled.

For the reaction:

$2\,NO_2(g) \rightarrow 2NO(g) + O_2(g)$, the rate law is Rate = $k[NO_2]^2$.

That is, the rate of the decomposition of NO_2 is *second order* with respect to $[NO_2]$. Therefore, the rate will quadruple if $[NO_2]$ is doubled.

The reaction: $2\,HI(g) \rightarrow H_2(g) + I_2(g)$ is found to be a zero order reaction in the presence of gold catalyst. For this reaction, the rate is independent of [HI].

Determination of the Rate Law

Method of Initial Rates

For many chemical reactions, the expression of rate law contains only the concentrations of reactants, while the concentrations of products do not appear. However, for some reactions that are reversible, the concentration of products may be part of the rate law. To simplify the rate law and avoid the complication that might be contributed by products in reversible reactions, the measurement of rate for rate law determination is normally done at the moment the reactants are mixed.

At this point, we can also assume that the effect of product concentration is considered negligible. Therefore, the rates that are measured are the *initial rates*. The Graphical Method for determining Zero-order, First-order and Second-order Rates

Consider a reaction: R $\rightarrow$ Products.

- If the reaction is *zero-order* with respect to R, Rate $= \dfrac{-\Delta[R]}{\Delta t} = k$

This gives D[R] = $-k\Delta t$, and $[R]_t = [R]_0 - kt$; a plot of [R] *versus* t yields a straight line with slope = $-k$

- If the rate is *first-order* in [R], Rate $= \dfrac{-\Delta[R]}{\Delta t} = k[R]$, which also gives

$$\frac{\Delta[R]}{[R]} = -kDt,$$

Integrating this equation from time t = 0 to t = *t*, yields the expression: $ln[R]_t = ln[R]_o - kt$; or $ln([R]_t/[R]_o) = -kt$; If the common log is used, the integrated expression for first rate would be:

$\log[R]_t = \log[R]_o - kt/2.30$; where, $[R]_o$ = concentration at t = 0, and $[R]_t$ = concentration at time *t*.

Plotting $ln[R]_t$ *versus* t yields a straight line with slope = $-k$; (slope = $-k/2.30$ if $\log[R]_t$ is plotted).

- If the rate is *second-order* with respect to R, Rate $= \dfrac{-\Delta[R]}{\Delta t} = k[R]^2$

$$\text{or } \frac{\Delta[R]}{[R]^2} = k\Delta t\,;$$

Integrating the equation from time t = 0 to t = *t* yields the expression:

$$\frac{1}{[R]_t} = kt + \frac{1}{[R]_0}$$

For second order rate, plotting $1/[R]_t$ *versus* t yields a straight line with the slope = k (the *rate constant*)

Half-Life of Reactions

The *half-life* ($t_{1/2}$) of a reaction is the time it takes for the reactant concentration to decrease to one-half its initial value.

For *zero-order* reaction, $t_{1/2} = [R]_0/2k$

For *first-order* reaction, $t_{1/2} = 0.693/k$

For *second-order* reactions, $t_{1/2} = 1/(k[R]_o)$

Reactions with More Than One Reactant

For reactions that involve more than one reactant, the integrated rate law is determined for one reactant at a time. This is done by having the concentration of the limiting reactant (which rate order to be determined) considerably lower than that of the other reactant. Under this condition, the time dependent change in concentration is significant only for the limiting reactant, while the other reactant is assumed approximately constant.

For example, the reaction: $2NO(g) + O_2(g) \rightarrow 2NO_2(g)$ has Rate = $k[NO]^2[O_2]$ The reaction can be carried out using $[NO]_o = 1.0 \times 10^{-4}$ mol/L and $[O_2] = 0.10$ mol/L.

As the reaction progress, [NO] decreases significantly, but the decrease in $[O_2]$ is negligible. Under this condition, the integrated rate law equation can be derived in term of [NO].

Then condition can be reversed and the integrated rate law equation is derived in term of $[O_2]$.

For this reaction, a straight line with positive slope is obtained for the plot of 1/[NO] *versus* time and for $ln[O_2]$ *versus* time, which yield a rate law: Rate = $k[NO]^2[O_2]$.

Reaction Mechanism

A *reaction mechanism* is the detailed picture of how a reaction occurs at molecular level.

It consists of *a set of proposed elementary steps* involving molecular species – reactants as well as reaction *intermediates*.

A reaction mechanism explains how a given reaction might occur at molecular level and from which a rate law can be derived, which must agree with the one determined experimentally. If the mechanism consists of more than one *elementary step*, the sum of these *steps* must be equal to the overall balanced equation for the reaction.

For example, the reaction: $2A + B \rightarrow C + D$ may involve the following *elementary steps*:

- *Step-1:* $A + B \rightarrow X$;
- *Step-2:* $X + A \rightarrow Y$;
- *Step-3:* $Y \rightarrow C + D$
- *Overall reaction:* $2A + B \rightarrow C + D$;

Elementary Steps and Molecularity

Elementary steps or *elementary reactions* are *simple* steps that together make up the *reaction mechanism* for a given reaction. Each *elementary reaction* describes *individual molecular event*, such as two particles combining or a particle (simple or complex) decomposing.

An *elementary reaction* is characterised by its *molecularity*, which is the number of particles/molecules involved in the formation of *transition-state complex*. Thus, an elementary reaction may be characterised as *unimolecular*, *bimolecular*, or *termolecular*.

Unimolecular and bimolecular reactions are considered the most common elementary processes.

Rate Law for Elementary Reaction Steps:

Elementary Reactions	*Molecularity*	*Rate Law*
A → product	Unimolecular	Rate = k[A]
2A → product	Bimolecular	Rate = k[A]^2
A + B → product	Bimolecular	Rate = k[A][B]
2A + B → product	Termolecular	Rate = k[A]^2[B]

For example, in the reaction: $2N_2O_5(g)$ ® $4NO_2(g) + O_2(g)$, the proposed mechanism is:

Step-1: $N_2O_5 \rightarrow NO_2 + NO_3$ (*Unimolecular*)
Step-2: $NO_2 + NO_3 \rightarrow 2NO_2 + O$ (*Bimolecular*)
Step-3: $N_2O_5 + O$ $2NO_2 + O_2$ (*Bimolecular*)

Termolecular event is considered very rare; the chances that three particles colliding at the same time, with proper orientation and sufficient energy, are considered extremely small. A possible *termolecular* event is thought to occur during the formation of ozone from oxygen in the outer atmosphere:

$$2\,O_2 + N_2 \rightarrow O_3 + O + N_2^*$$

Deriving Rate Law from the Rate-Determining Step

When a proposed mechanism consists of more than one elementary steps, the one with the slowest rate will determine the overall rate of reaction. It is called *the rate-determining step*, which *step represents the rate law for the overall reaction.*

Consider the reaction: $NO_{2(g)} + CO_{(g)} \rightarrow NO_{(g)} + CO_{2(g)}$

If the reaction follows a *one elementary step* mechanism, the rate law would be: Rate = $k[NO_2][CO]$

However, the experimentally determined rate law is Rate = $k[NO_2]^2$

It is proposed that the reaction must involve at least *two elementary steps*:

$$\text{step}-1: NO_2 + NO_2 \xrightarrow{k_1} NO_3 + NO; \left[\text{slow; rate}-\text{determining}\right]$$

$$\text{step}-2: NO_3 + CO \rightarrow CO_2 + NO; \quad \left[\text{fast}\right]$$

The rate law for the rate-determining step: *Rate* = $k_1[NO_2]^2$, which is identical in form to the rate law obtained experimentally, in which $k_1 = k$. The second step, which occurs very fast, does not influence the overall rate.

In fact, if the exponents in the rate law differ from the coefficients in the chemical equation, the mechanism for the reaction normally involves more than one elementary step. Most reactions are believed to involve more than one elementary step. Note that, unlike the overall rate law which must be determined experimentally, the rate law for an *elementary reaction* can be deduced from its *molecularity*, since each elementary reaction is a single step event. The exponents in the expression of rate law are the same as the equation coefficients. For example, the rate laws for the following elementary reactions are written as indicated:

1. $NO + O_3 \rightarrow NO_2 + O_2$; Rate = $k[NO][O_3]$
2. $N_2O_5 \rightarrow NO_3 + NO_2$; Rate = $k[N_2O_5]$
3. $Br_2 + NO \rightarrow ONBr_2$; Rare = $k[Br_2][NO]$

In addition, certain species such as NO_3 may be formed in on *elementary step*, but immediately consumed in subsequent elementary step; they do not appear in the overall equation. These species are called *reaction intermediates.*

Rate Law Derived from Mechanism with Slow First Elementary Step

Consider the reaction:$2NO_2(g) + F_2(g) \rightarrow 2NO_2F(g)$

The experimental rate law is Rate = $k[NO_2][F_2]$ The proposed mechanism for this reaction consists of the following elementary steps:

Step-1: $NO_2 + F_2 \rightarrow NO_2F + F$; (slow, rate-determining)

Step-2: $NO_2 + F \rightarrow NO_2F$; (fast)

Overall Rxn: $2NO_2 + F_2 \rightarrow 2NO_2F$

Rate law for the rate-determining step:

Rate = $k_1[NO_2][F_2] = k[NO_2][F_2]; \rightarrow (k = k_1)$

Rate Law Derived from Mechanism with *Slow Second Elementary Step*:

Consider the reaction: $2\ NO(g) + O_2(g) \rightarrow 2\ NO_2(g)$;

The rate law is: Rate = $k[NO]^2[O_2]$

The proposed mechanism composed of the following elementary reactions:

Step-1: $NO + O_2 \rightleftarrows NO_3$; (fast; equilibrium)

Step-2: $NO_3 + NO \rightarrow 2\ NO_2$; (slow, rate-determining)

Overall Rxn:$2NO + O_2 \rightarrow 2\ NO_2$;

Rate law for the *slow, rate-determining step*: Rate = $k_2[NO_3][NO]$

NO_3 is a *reaction intermediate* formed in a fast reversible reaction (step-1) and it concentration quickly reaches a steady state such that, $[NO_3] = K_{eq}[NO][O_2]$, where K_{eq} is the equilibrium constant for reversible step-1. Substituting for $[NO_3]$ in the rate law expression, we obtain,

Rate = $k_2K_{eq}[NO]^2[O_2] = k[NO_2]^2[O_2]$;

(where $k = k_2K_{eq}$)

Correlating Reaction Mechanism with Rate Law

For any reaction, the proposed *mechanism* must obey three fundamental criteria:

1. *The elementary steps must add up to give the overall equation.*

2. *The elementary steps must be physically reasonable*. That is, the proposed step can occur with reasonable probability; they are either *unimolecular* or *bimolecular*. A *termolecular* elementary step is considered unlikely.
3. *The mechanism must correlate with the rate law*. That is, the rate law derived from *rate-determining step* must agree with the actual rate law obtained experimentally.

For example, the reaction:

$2\ N_2O_5(g) \rightarrow 4\ NO_2(g) + O_2(g)$ has Rate = $k[N_2O_5]$.

The proposed mechanism involves the following elementary steps:

Step-1: $N_2O_5 \rightleftarrows NO_3 + NO_2$; (fast; equilibrium)

Step-2: $NO_3 + NO_2 \rightarrow 2NO_2 + O$; (slow)

Step-3: $N_2O_5 + O \rightarrow 2NO_2 + O_2$; (fast)

Overall: $2\ N_2O_5 \rightarrow 2NO_2 + O_2$; Rate = $k[N_2O_5]$

A reproduction for Chemical Kinetics

The rate law of a reaction relates the dependence of reaction rates on the concentrations of reactants. As we know, the rate of reaction also depends on other factors, such as temperature.

This dependency can be explained by the *collisional theory* of reaction and the formation of *activated complex*, or *transition state*. Consider the following reaction: $2BrNO(g) \rightarrow 2NO(g) + Br_2(g)$

For the reaction to occur:

- *Reactant molecules must collide;*
- *Molecular collisions must occur with proper orientations;*
- *Collisions must be energetic and lead to the formation of an activated complex;*
- *The rate of formation of activated complex is the rate determining step;*
- *The activated complex eventually leads to the formation of products;*

1. The rate of formation of *activated complex* is proportional to the *frequency of effective molecular collisions*, which is dependent on reactant concentrations.
2. The *activated complex* or *transition state* is separated from the reactants by an *energy barrier*, called the *activation energy* (E_a). The formation of *transition state complex* depends on the fraction of molecules with sufficient *kinetic energy* that can overcome this *energy barrier*. This fraction is temperature dependent. At low temperature, the *fraction* of molecules with sufficient energy to overcome the *activation energy barrier* and subsequently forms the *transition state complex* will be low. This *fraction* increases with temperature, which also increase the overall reaction rate.

3. High *activation energy* for a reaction also means a lower fraction of molecules with sufficient energy to form the *transition state complex*. Thus, higher E_a leads to a slower rate of reaction.

The dependence of rate on temperature and *activation energy* (E_a) is shown the Arrhenius equation for the rate constant, *k*:

$k = Ae^{-Ea/RT}$, where *A* is called the *Arrhenius frequency factor*.

Taking the natural log, yields the equation:

$$\text{In}(k) = -\left(\frac{E_a}{R}\right)\left(\frac{1}{T}\right) + \text{In}(A).$$

A plot of *ln*(*k*) *versus* 1/T yields a straight line with slope = $-E_a/R$; (R = 8.314 J/mol.K).

The *activation energy* (E_a) can also be calculated using two values of the rate constant *k* according to the expression:

$$\text{In}(k_2/k_1) = -\frac{E_a}{R}\left(\frac{t}{T_2} - \frac{1}{T_1}\right); \text{ or } \text{In}(k_1/k_2) - \frac{E_a}{R}\left(\frac{1}{T_2} - \frac{1}{T_1}\right);$$

(T must be in Kelvin)

Catalysis increase the rate of reactions

Catalysts are substances that increase the rate of reactions, but do not get used up by the reactions. A catalyst functions by providing an *alternative reaction pathway with lower activation energy*. It increases the reaction rate, but does not affect the reaction enthalpy or the equilibrium position.

Homogeneous Catalysis

These are catalysts that have the same phase as the reactants. For example, nitric oxide (NO) catalyses the reaction between SO_2 and O_2:

$$2SO_2(g) + O_2(g) \xrightarrow{\text{NO catalyst}} 2SO_3(g)$$

The above reaction, though spontaneously and highly exothermic, is very slow because of the very high activation energy. NO molecules provide an alternative pathway that has a lower E_a.

Step-1: $2NO + O_2 \rightarrow 2NO_2$

Step-2: $2NO_2 + 2SO_2 \rightarrow 2NO + 2SO_3$

Overall: $2\ SO_2(g) + O_2(g) \rightarrow 2SO_3(g)$

Some catalysts may be involved in the reaction kinetics and the concentration appears in the rate law. For example, iodine which catalyses the *isomerisation* of *cis*-2-butene to *trans*-2-butene is included in the rate law expression. The Uncatalysed reaction:

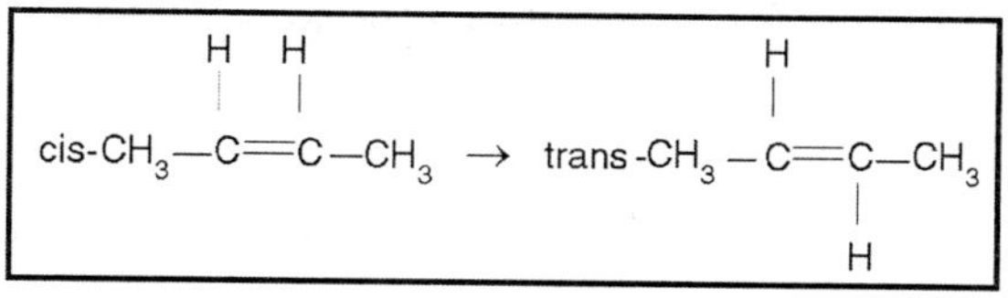

Rate = $k[cis\text{-2-butene}]$

Iodine-catalysed reaction:

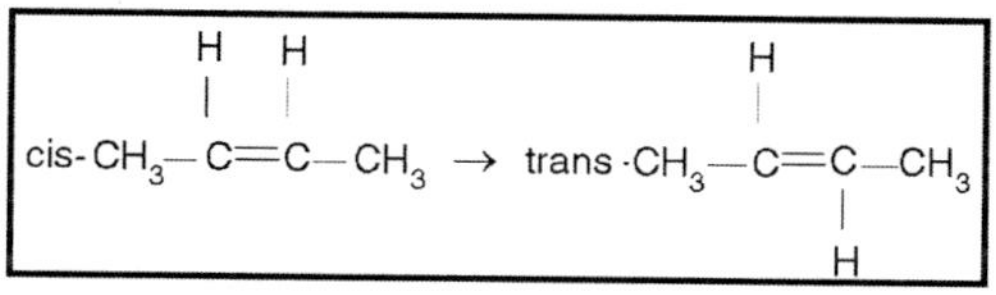

Rate = $k[cis\text{-2-butene}][I_2]^{1/2}$

Heterogeneous Catalysis

These are catalysts which have a different physical state from the reactants. Most reactions involving gases use inert metals or metal oxides as catalysts. These solid catalysts provide surface areas for molecules to interact effectively. The metal surface also facilitates the breaking and formation of bonds. Ni, Pd and Pt are often used in the hydrogenation of vegetable oil to make margarine and "Crisco oil", where a hydrogen molecule is added to each carbon-carbon double bond in the unsaturated fatty acid chains:

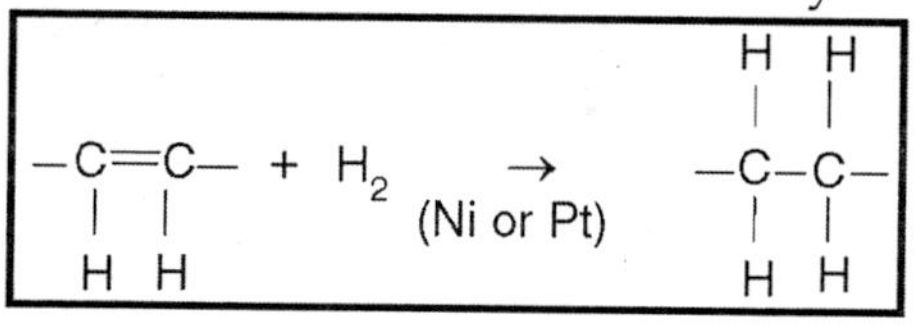

Heterogeneous catalyses are used in automobile catalytic converters, which consist of Pt, Pd and Rh mixture embedded in ceramic honeycombs. The metals catalyse the following reaction:

$$2NO(g) + 2CO(g) \xrightarrow{(Pt+Pd+Rh)} N_2(g) + 2CO_2(g)$$

Catalytic converters reduce the emission of toxic exhaust gases

Some reactions require specific catalysts. For example, Nickel catalyses the reaction:

$$CO(g) + 3H_2(g) \xrightarrow{\text{Ni-catalyst}} CH_4(g) + H_2O(g)$$

Whereas ZnO-Cr_2O_3 mixture catalyses the formation of methanol from the same reactants:

$$CO(g) + 2H_2(g) \xrightarrow{ZnO-Cr_2O_3} CH_3OH(g)$$

2

Structure of an Atom

An atom is composed of three different types of particles; protons, neutrons, and electrons. Because atoms are neutrally charged, an atom will always have the same number of protons and electrons. This idea that atoms are neutral is by definition. Atoms always have the same number of negative and positive charges.

By the way, atoms become ions when they become charged. This change can occur when an atom has either lost or gained some electrons. So ions do not have the same number of protons and electrons.

The example atom to the left or above, represents Helium which is usually a gas. Helium has 2 protons, 2 neutrons, and 2 electrons. The protons and neutrons form the nucleus and the electrons form a cloud that surrounds the nucleus. Now what is it about this atom that makes it a Helium atom? It is the number of protons that determines what kind of element the atom will be. A Helium atom will always have 2 protons. If it is a non-charged atom, it will also have 2 electrons.

There are; However, several varieties of Helium and they differ in the number of neutrons. Scientists call these varieties isotopes. Most of the Helium that we breath in our air is Helium 4 which has 2 neutrons. There are also a small number of Helium 3 atoms in our air as well. Both of these varieties of Helium are stable. They do not break down. So they are not radioactive.

EVALUATION OF THE ATOMIC CONCEPT

Dalton first recognized and effectively utilized the stimulating force of the atomic concept and its ability to guide future physical research. To him we owe the use of the idea of atoms for understanding basic regularities of chemistry. Chemistry makes a sharp distinction between mixtures of different substances and chemical combinations. A mixture–as, for example, a solution of sugar in water or a mixture of nitrogen and oxygen–can include arbitrary (within certain limits) relative proportions of the constituents. In a nitrogen-oxygen mixture the portion of oxygen can be varied continuously by addition or removal; correspondingly, the mixture exhibits properties somewhere between those of pure oxygen and of pure nitrogen, depending upon the relative concentration.

Also, the mixture can be separated into its constituents through relatively superficial means. In a chemical combination entirely new properties appear, which can be quite dissimilar to those of the constituents. The proportions of the components of a chemical compound are quite stable and cannot be varied continuously.

Thus exactly 16 grams of oxygen always combine with 2.016 grams of hydrogen to form water. Another chemical substance, hydrogen peroxide, can be formed from 2.016 grams of hydrogen and just 32 grams of oxygen. It is striking that exactly twice as much oxygen is utilized in the second case as in the first; and on the basis of the atomic concept Dalton found a clear explanation for such facts, which occur similarly in all chemical combinations. His explanation was as follows: The "molecule" of water, *i.e.*, the smallest possible particle of water, contains only one atom of oxygen; the molecule of hydrogen peroxide however contains two atoms of oxygen, while it contains just as many (two) atoms of hydrogen as the water molecule.

Here we arrive at the distinction between atoms and molecules; we designate the smallest particles of chemical compounds as molecules, which for their part consist of atoms of the chemical elements. According to Dalton's ways of thinking a comprehensive study of proportions by weight in chemical compounds rendered the determination of the relative atomic weights of all the elements possible. The "atomic" weight of oxygen is arbitrarily designated by the value 16, whereupon hydrogen acquires the atomic weight 1.008 and every other element, similarly, receives a definite atomic weight derived from chemical weight relationships.

The regularities which the gases exhibit, which in the case of sufficiently small densities assume an "ideal" form, proved very helpful for the clarification of these ideas; for example, in the sense that a gas mass contracts to one-half its original volume when, with constant temperature, the pressure impressed upon it is doubled.

Criteria existed for the hypothesis that not only in compounds but also in a gaseous element such as nitrogen the particles moving at random through space are not necessarily identical with atoms; in nitrogen, for example, each "molecule" of the gas consists of two atoms. In the chemical combination of gases (with constant pressure and temperature) there is a simple relation between proportions by weight and volume relations (proportions by volume) between the chemical compound and the still uncombined constituents.

To fit these facts Avogadro introduced the explanation that equal volumes of all gases measured at constant pressure and temperature always contain the same number of molecules. Based on this "Avogadro's principle" a "molecular weight" can be determined for each gas independent of chemical weight relations.

The molecular weight of a chemically homogeneous gas is equal to the mass of 22.4 litres of this gas measured in grams at one atmosphere of pressure and a temperature of 0° C. Experience shows that values obtained in this way correspond with the chemically defined atomic weights. In many elements, *e.g.*, the metals, this molecular weight is exactly the same as the atomic weight; thus metallic vapours are "monatomic". In the aforementioned nitrogen, however, the molecular weight is exactly double the atomic weight. And in water it equals 18.016 ($2 \times 1.008 + 16$) in accordance with its above-described chemical composition.

These ideas are further verified by simple relations of molecular weights such as, "diffusion velocities", or by the characteristic difference between the specific heats of "monatomic" and "polyatomic" gases. None of these, it must be emphasized, are proofs for the correctness of the atomic concept; but they are proofs for its usefulness. The atomic idea describes very plainly a large number of important and comprehensive regularities, thus fadlitating greatly our progress with these phenomena.

Crystallography furnishes evidence of further regularities which are certainly clearly explained by the atomic idea, though they too do not lead to proofs for the reality of the atom. In the crystalline state of matter the atomic concept suggests a regular grouping of the atoms or molecules, alongside of each other and arranged in layers.

This view induces important results; on this basis the variety of possible forms of crystals are shown mathematically to be limited very considerably. The completion of this mathematical proposition yields the wonderful result that according to the atomic concept there should be no more nor less than 32 different "crystal classes" which are defined by different properties of symmetry.

Actually this confirms precisely the experience of mineralogists; there really occur in nature all the crystal forms (crystal symmetries) which are compatible with the atom idea, but not one single one which contradicts it. The so-called "law of rational indices" discovered by the mineralogists formed an important adjunct to this original consideration, which is related only to the symmetry properties of different crystal forms. If we imagine the crystal as built up of a regular stratification of layers of atoms, then its outer boundary surfaces could never lie so that they partially cut through the atoms. The limits for the possible positions of the boundary surfaces given by this view are exactly equivalent to the requirements which the law of rational indices places on the position of the crystal surfaces.

Such experiences must necessarily encourage the more and more energetic pursuit of the atomistic notion–which intrinsically was so clear–even though tangible proof for the reality of atoms was still lacking. Actually the theoretical development of the atomic concept was furthered extensively by the most able physicists. We have already mentioned ideal gases and indicated how the gaseous state of aggregation should be represented on the basis of the atomic theory; the molecules of a gas are completely separate from each other; each one moves along with high velocity until it meets another one, and then, in elastic rebound, their direction of motion and speed change.

The continuous impact of countless molecules on the walls of the enclosing vessel produces the pressure of the gas, perceptible to us with our rough tools. These ideas were elaborated thoroughly (Clausius, Maxwell) and our understanding of the gaseous state of aggregation was developed very clearly and completely. Whereby the proof resulted that Avogadro's principle, first only speculatively suspected, must necessarily be correct if the atomic hypothesis proves at all valid.

The discovery of the energy principle (Mayer, Joule) further strengthened confidence in atomism. It was recognized that mechanical energy, which disappears through friction or the like, reappears in the quantity of heat produced–in such a way that a definite amount of consumed mechanical energy corresponds to a definite number of calories of heat.

Conversely, according to the same conversion relation, in steam engines quantities of heat are transformed again into mechanical kinetic energy. This conversion of mechanical energy into heat and its converse can be considered quite consistent, in the sense of "macrophysics". Whereupon one must establish that energy, which has proved indestructible, can assume both the form of a quantity of heat and of mechanical energy. The "macrophysical", "phenomenologic" heat theory thus arrived at is entirely sufficient for answering all the questions encountered in connection with heat engines, and with heat conversion in chemical processes.

But further thought is suggested to give the process of the conversion of mechanical energy into heat a clear interpretation in the light of the atomic theory. In a gas, as was represented above, the concepts of heat and temperature are not applicable to the individual molecules, but only to the gas mass as a whole. In an individual molecule only the kinetic energy, *i.e.*, its velocity of motion, or at best also its rotation, can be changed.

Thus, if energy is added to a gas mass, which appears macrophysically as added heat, it can only mean that the average energy of motion of the gas molecules has increased. The exact determination of the connection between the average energy of motion of the gas molecules and the temperature of the gas is a part of the theoretical investigations on the "kinetic theory of gases", which has already been discussed.

Similarly, in a solid body, a crystal perhaps, the temperature and heat energy contained within it are conceived of as mechanical energy of its molecules or atoms. Although in their regular arrangement in layers no one atom can leave its place, fine vibrating motions within the crystal–possibly similar to the fine vibration of a heavily travelled steel bridge–are still possible.

If these vibrations within the crystal are so small that they are not perceptible to us as mechanical motion, we do note the energy of these motions–as heat content which manifests itself in the temperature of the body. (If the atoms contained in each small piece of the crystal move by very small amounts in random directions, the whole crystal appears motionless to our macrophysical senses.) Apropos of such thought processes was Boltzmann's atomistic interpretation of a remarkable regularity in heat phenomena. If we imagine that a planet's motion is interrupted and the planet then is induced to move in the opposite direction with the same velocity, it will retrace its entire elliptical path in the opposite direction after the reversal.

This is an example of the fact that all purely mechanical motions are, as we say, reversible. The upward motion of a stone, thrown from the earth, (in the ideal case) until its-point of reversal is an exact temporal mirror image of the subsequent downward motion. Similar reversibility–which can also be denoted as a physical "symmetry of the positive and negative time orientation"–exists in all purely electromagnetic properties. But when we move a body over a table surface against frictional forces and heat is generated thereby, there is no reason to expect that in moving the body backwards along the same line the converse is true–that the heat energy will change back to energy of mechanical motion. Or take an electric current which flows through a wire and produces heat–when it flows through the wire in the opposite direction it again produces heat; the heat energy does not change back into electrical. These are examples of irreversible processes. Another example of such a process is the mixing of two liquids; in most cases these intermix by themselves, but this process never runs backward of itself.

A further example of importance to us is the following: if a gas filled vessel is introduced into an evacuated one and is opened there, the gas diffuses throughout the entire available volume. It is not impossible to restore the original conditions; the larger vessel can be evacuated again by pumping and the gas can be compressed into the smaller one.

But this does not constitute an exact reversal–a temporal reflected image–of the first process; and it can be considered that for the restoration of original conditions after any irreversible process a definite "compensation" must be included (Clausius). In the steam engine, which undertakes to convert quantities of heat into mechanical energy, there does not occur a simple reversal of the process of the production of heat through friction, but each steam engine functions according to a scheme similar to the following: a certain amount of heat is removed from a heat container maintained at a high temperature. A fraction of this amount is changed into mechanical work, while the remainder is conducted into a heat reservoir of lower temperature. As "compensation", therefore, for the conversion of heat into work there occurs the transfer of a quantity of heat from a hotter body to a colder one–*i.e.*, therefore a process which nature could perform "irreversibly" itself, since in heat conduction the heat always flows from higher to lower temperature.

The occurrence of such "irreversible" processes among heat phenomena posed a very difficult problem for the "kinetic theory of heat"–tracing back the laws of heat to the mechanics of atoms. For purely mechanical processes were shown to be always reversible, which makes it appear impossible to trace these irreversible heat phenomena back to mechanical properties.

The solution of this difficulty, achieved through Boltzmann's perspicacity, runs as follows: in an aggregation of many similar particles, such as are attributed to every physical system by the atomic theory, there is no point to pursuing the motion of each individual atom or molecule; it is only important to us to observe the average statistical behaviour of large numbers of atoms.

Thus statistical concepts are necessarily introduced into the consideration and we must determine which (defined in a rough statistical sense) events are to be considered as especially probable (thus as occurring very frequently) in comparison with other conceivable processes.

For our example this signifies that when the small gas filled vessel is opened in the large empty one "it is extremely probable" that the gas will stream out and distribute itself uniformly in the large vessel. Strictly speaking it cannot be definitely known that this will happen; indeed, to view the processes of motion in the gas mass with the same complete certainty achieved for planetary motions one must know quite exactly the position and velocity of each individual molecule at the beginning of the experiment. The fact that, instead of this, we simply denoted the initial condition of the experiment statistically stipulates that we can not predict its further progress with complete certainty, but only with very great probability.

The reversibility of pure physical processes, which according to the atomistic conception is the basis for macrophysical heat phenomena, depends upon the existence of an exact reversal, an exact temporal reflected image for every process–whereby it can happen that the gas mass distributed in the above mentioned large vessel contracts into the smaller one, in an exact reversal of the normal process. No one can guarantee that this is completely out of the question; but we can show that such an event is extremely improbable–mathematically its probability is expressed by an infinitesimally small number. Thus it becomes clear that despite the fundamental reversibility of atomistic elementary-processes, yet, on a large scale, in macrophysical events, practically irreversible phenomena take place. Another instructive example of an irreversible process is the mixing of two gases or liquids with each other. According to the kinetic-atomic theory this problem is somewhat similar to the one wherein two types of balls–red and white ones for instance–are thoroughly shaken up in a sack. If the red and white spheres are carefully separated initially, they will become completely mixed up after further agitation. If, then, we continue to shake, it is possible in principle that the original separation of red and white balls will be restored–but practically this possibility is of little moment since it would require shaking for astronomically long periods of time before this "spontaneous separation" could be expected with appreciable probability.

A quantitative measure for the irreversibility of a thermodynamic process (from a purely macrophysical standpoint) can be specified in the form of "entropy"–a quantity which always increases to a maximum in irreversible processes and does not decrease again (Clausius). According to Boltzmann this entropy can now be defined through the atomic concept as a measure of the disorder in an atomic aggregate.

ELECTRON GROUPING IN ATOMS

Before proceeding with the description of the more common periodic tables it seems appropriate to list the types of atoms which are recognized in all schemes of classification. This listing emphasizes the fact that the chemical properties of the elements depend upon the extent to which their electron shells are completed.

By the completion of a shell we mean the agreement between the number of electrons it contains and the number in the corresponding shell of the noble gas of the next higher or lower atomic number. The symbolic representation of various electronic configurations. Upon the basis of complete and incomplete electron shells four types of atoms are recognized.

Type I. Atoms in which the s and p sublevels are complete. The Pauli exclusion principle predicts that the s level can accommodate not more than 2 electrons and the p level not more than 6 electrons. When these levels are complete, the outer shell has a configuration of ns2np6, except for helium, in which the configuration is 1s2.

These are the atoms of the noble, or inert, gas elements, which have attained an electronic stability which cannot be increased either by electron transfer or by sharing. Because of the stability of their electronic configurations such elements do not enter into ordinary chemical reactions. These elements make up Group 0 of the periodic table.

Type II. Atoms in which all but the highest quantum group (outermost shell) are complete. The electronic configurations in the outer shells of atoms of these elements include any structures from ns1 through ns^2np^5. A comparison of tablewhich shows the distribution of electrons in the atoms of the elements, with the Mendeleev periodic grouping as shown in tableindicates a relationship between the quantum mechanically permitted orbitals and the placement of elements in the periodic table.

For example, the elements of Group IA have an incomplete s level in the outer electronic shell, the elements of Group IIA have a completed s level, and the atoms of the elements of Groups IIIB to VIIB have incomplete p levels in the highest quantum group.

Elements of this type of atom tend to enter into chemical combination through losing, gaining, or sharing electrons, in such wise as to attain the electronic configuration of ns^2np^6 for the outermost shell, which is the same as that found in the noble gases. Many of the elements most often encountered belong to this type. Because of the frequency of their occurrence they are sometimes called the "representative" elements, although such a designation is more or less of a misnomer.

Type III. Atoms in which the *two* outermost shells are incomplete. They are atoms of those elements in which an inner d level is present but not completed. The electronic configurations for the outer shells of these elements include those with a general structure of $(n-1)d^{1-9}ns^{1-2}$. In other words, the next to the outer shell has a *d* level which may contain one through nine electrons, and the outer shell contains either one or two electrons.

Elements with atoms of this general structure have been named *transition* elements. The transition series occurring among the elements are those with incomplete *3d, 4d, 5d,* and perhaps *6d* orbitals. In the periodic table these elements begin with Group IIIA and continue through Group IIB.

However, the elements of Group IB (the copper family) and Group IIB (the zinc family) have electronic structures similar to the Type II atoms, but in terms of chemical behaviour they have characteristics that are analogous to the transition elements. Because of their electronic structure the transition elements exhibit variable valence.

Type IV. The atoms of elements of this type are actually transition elements, but they differ from atoms of the regular transition elements in that they have *three* outermost shells which are incomplete. They may be more properly identified as *inner transition elements,* inasmuch as they not only have incomplete *d* levels but also *f* levels that have not been filled to capacity. These are the elements of the rare earth, or lanthanide, series (atomic numbers 58 to 71) and those of the actinide series (atomic numbers 90 to 102).

At present it suffices to say that the elements within each of these groups are very similar in their chemical properties because the differentiating electrons of their atoms are in the third from the outermost shell. Because of the similarity of chemical behaviour among the elements of the two inner transition series they cannot be accommodated in the conventional periodic tables. Hence it is customary to indicate the two inner transition series as separate groups of elements.

The Modern Mendeleev Periodic Table

It is quite likely that too much credit has been given to Mendeleev for the development of the first widely accepted periodic table, but there is no doubt that the table attributed to him has been the forerunner of all other worthwhile systems of classification for the elements. Acknowledgedly it has many faults, some of which we shall point out later, but so have all periodic tables.

Furthermore the majority of modern classifications are the results of attempts to correct these recognized faults. On the other hand the Mendeleev table is superior to all other tables in one respect, and that is its simplicity. The sole purpose of a periodic table is as a valuable tool in the teaching of chemistry. Undoubtedly other forms of the periodic table are more meaningful, but for a beginning student it is probable that the Mendeleev table will always be of value.

A modern version of Mendeleev's periodic table is given in table. An examination of this table discloses that it is divided horizontally into *periods* — rows across the table from left to right — whereas, vertically, it is divided into columns called *groups*. There are in all seven periods, of which the first is unique in that it is occupied by only one element, hydrogen.

Each of the next two periods contains 8 elements, arranged in a single row called a *short period,* across the table. Following these are two *long periods,* containing 18 elements each, arranged in the form of two rows across the table. The sixth is a very *long period,* consisting of 32 elements, 15 of which are the so-called "rare earths." The seventh and last period appears incomplete, inasmuch as it contains the actinide series of elements, all of which, as yet, have not been discovered.

The vertical columns are occupied by elements with related chemical properties. With the exception of Groups 0 and VIII each column contains two subgroups, or families, usually designated as A and B. The elements within a subgroup resemble one another more closely than they resemble the members of the other family subgroup.

Thus, in Group I, Li, Na, K, Rb, and Cs are very closely related chemically and constitute family A, whereas Cu, Ag, and Au make up family B, the members of which, although similar to each other, do not bear close resemblance to the elements of family A. Useful as the modern Mendeleev classification may be in presenting many facts of chemistry in a convenient, concentrated form, it contains many inherent defects.

One of the first faults which became evident at an early date was the inversion among certain atomic weights. The periodic law was originally stated by Mendeleev in terms of atomic weights. Later the investigations of Moseley showed that the properties of the elements are periodic functions of their atomic numbers.

And not atomic weights. Moseley's work, therefore, explained the inversions in atomic weights occurring in argon and potassium, in cobalt and nickel, and in tellurium and iodine. Other defects which have not been "ironed out" as satisfactorily are described in the following paragraphs.

The properties of the element hydrogen are so unique that it does not fit into any of the eight groups of Mendeleev's periodic table. Because it exhibits a positive valence toward oxygen of + 1, hydrogen is frequently placed in Group I, although its nonmetallic properties, and its ability to form a negative ion, would indicate that it might fit just as well, or even better, in Group VII. An unsatisfactory compromise is to place it in both Groups I and VII.

The Mendeleev-periodic table indicates relations that do not necessarily exist. Except for common valences there is usually very little resemblance between the members of the two families in a given group. Certainly there is little chemical similarity between the alkali metals and copper, silver, and gold in Group I. Furthermore even within a family the classification may be somewhat artificial. For example, bivalent copper could fit into family B of Group II, as well as or better than monovalent copper fits into family B of Group I.

Two groups of elements, the lanthanum series of rare earths and the actinium series of radioactive elements, do not fit into the Mendeleev classification.

To place either or both of these groups into the body of the table would destroy the effectiveness of the system of classification. For that reason these elements are omitted from the main part of the table and are placed, instead, in separate rows beneath it.

The table places undue emphasis upon one valence of each element, and frequently the valence that is stressed is not the most important one. For instance, copper is indicated by the table as having a valence of 1, which is true; however, copper also has a more frequently occurring valence of 2 that is not indicated by the table.

Other defects of the Mendeleev periodic table may be briefly stated:

1. Three groups of elements, each group consisting of three members, for example, Fe, Co, and Ni, must be placed in Group VIII, although these elements rarely if ever exhibit a valence of 8.
2. No attempt is made to indicate a separation of metals from nonmetals.
3. Despite the fact that chemical properties are directly related to the electronic configuration of various atoms, no satisfactory correlation of this relationship is provided in the table.

ELEMENTS OF THE FIRST TRANSITION SERIES

There are a number of properties shared by the transition elements that are not found in other elements, which results from the partially filled d shell.

These include:

- The formation of compounds whose colour is due to d - d electronic transitions,
- The formation of compounds in many oxidation states, due to the relatively low reactivity of unpaired d electrons,
- The formation of many paramagnetic compounds due to the presence of unpaired d electrons. A few compounds of main group elements are also paramagnetic *e.g.* nitric oxide, oxygen.

Coloured Compounds

Colour in transition-series metal compounds is generally due to electronic transitions of two principal types:

- *Charge transfer transitions*: An electron may jump from a predominantly ligand orbital to a predominantly metal orbital, giving rise to a ligand-to-metal charge-transfer LMCT transition. These can most easily occur when the metal is in a high oxidation state. For example, the colour of chromate, dichromate and permanganate ions is due to LMCT transitions. A metal-to ligand charge transfer MLCT transition will be most likely when the metal is in a low oxidation state and the ligand is easily oxidised. Mercuric iodide, HgI2, is red because of a MLCT transition. As this example shows, charge transfer transitions are not restricted to transition metals.
- *D-d transitions*: An electron jumps from one d-orbital to another. In complexes of the transition metals the d orbitals do not all have the same energy. The pattern of splitting of the d orbitals can be calculated

using crystal field theory. The extent of the splitting depends on the particular metal, its oxidation state and the nature of the ligands. The actual energy levels are shown on Tanabe-Sugano diagrams.

Oxidation States

A characteristic of transition metals is that they exhibit two or more oxidation states, usually differing by one. For example, compounds of vanadium are known in all oxidation states between –1, such as, and +5, such as.

Main group elements in groups 13 to 17 also exhibit multiple oxidation states. The "common" oxidation states of these elements typically differ by two. For example, compounds of gallium in oxidation states +1 and +3 exist in which there is a single gallium atom. No compound of GaII is known: any such compound would have an unpaired electron and would behave as a free radical and be destroyed rapidly.

The only compounds in which gallium has a formal oxidation state of +2 are dimeric compounds, such as $Ga_2Cl_62–$, which contain a Ga-Ga bond formed from the unpaired electron on each Ga atom. Thus the main difference in oxidation states, between transition elements and other elements is that oxidation states are known in which there is a single atom of the element and one or more unpaired electrons.

The maximum oxidation state in the first row transition metals is equal to the number of valence electron from titanium +4 up to manganese +7, but decreases in the later elements. In the second and third rows the maximum occurs with ruthenium and osmium +8. In compounds such as $MnO_4–$ and OsO_4 the elements achieve a stable octet by forming four covalent bonds. The lowest oxidation states are exhibited in such compounds as $CrCO_6$ oxidation state zero and $FeCO_42–$ oxidation state –2 in which the 18-electron rule is obeyed. These complexes are also covalent.

Ionic compounds are mostly formed with oxidation states +2 and +3. In aqueous solution the ions are hydrated by usually six water molecules arranged octahedrally.

Magnetism

Transition metal compounds are paramagnetic when they have one or more unpaired d electrons. In octahedral complexes with between four and seven d electrons both high spin and low spin states are possible. Tetrahedral transition metal complexes such as $FeCl_42–$ are high spin because the crystal field splitting is small so that the energy to be gained by virtue of the electrons being in lower energy orbitals is always less that the energy needed to pair up the spins.

Some compounds are diamagnetic. These include octahedral, low-spin, d6 and square-planar d8 complexes. In these cases, crystal field splitting is such that all the electrons are paired up.

Ferromagnetism occurs when individual atoms are paramagnetic and the spin vectors are aligned parallel to each other in a crystalline material. Metallic iron and the alloy alnico are examples of ferromagnetic materials involving transition metals. Antiferromagnetism is another example of a magnetic property arising from a particular alignment of individual spins in the solid state.

Other Properties

In general transition metals possess a high density and high melting points and boiling points. These properties are due to metallic bonding by delocalized d electrons, leading to cohesion which increases with the number of shared electrons.

However the group 12 metals have much lower melting and boiling points since their full d subshells prevent d-d bonding. In fact mercury has a melting point of –39 °C and is a liquid at room temperature. Some transition metals are useful in homogeneous and heterogeneous catalysis. For example, iron is a catalyst in the Haber process for ammonia synthesis, while nickel and platinum are used in hydrogenation of alkenes.

Transition Metals

The electronic configuration of these elements are characterised as having full outer orbitals and as having the second outermost orbitals incompletely filled. Thus, in the case of the first transition metals series, the electronic configuration of the outer orbitals is 4s2, while the second outermost orbitals (*i.e.* the 3d level) are incompletely filled.

Because they are elements whose atoms have their next-to-outer most main levels incomplete at the inner d sub-levels, but are filling up progressively, while the outer most level contains at most two electrons, the filling of the d orbitals is not always regular.

There are four transition series:

- *The first transition series*: Scandium (Sc) through Copper (Cu): 3d subshell is filling.
  Irregularities are observed for Chromium, Cr, and Copper, Cu, because the listed electronic configurations are energetically favoured (*i.e.* 3d54s1 is more stable than 3d44s2, and 3d104s1 is more stable than 3d94s1 respectively.).
- *The second transition series*: Yttrium (Y) through Silver (Ag): 4d subshell is filling.

Irregularities are observed for Nb which skips from 4d25s2 to 4d45s1, and Pd, which goes from 4d85s2 to 4d105s1.

- *The third transition series*: Lanthanum (La) to Hafnium (Hf) through Gold (Au)): 5d subshell is filling.
  Irregularities are observed for Pt which skips from 5d96s2 to 5d106s1.
- *The forth transition series which is incomplete*: Actinium (Ac) to element 104 through element 109: 6d subshell is filling, If elements 110 and 111 are found then this will complete this series..

Properties of Transition Metals

The Properties of Transition Metals are largely dependent on the electronic configuration of the electrons in the outer shell and in the penultimate outer shell.

The transition elements readily form alloys with themselves and with other elements (*e.g.* a copper-tin alloy is used for mirrors, brass is a copper-zinc alloy). Tungsten, is used to make tools and filaments in light bulbs. The atomic size is fairly constant since the electrons in the outer most shells have similar environments. The low ionisation potentials mean that the elements show variable valency states by loss of electrons from the s and 3d orbitals.

The elements in this group can have different oxidation states which makes them useful as catalysts. Compounds of the transition elements can be paramagnetic (*i.e.* attracted by a magnetic field) or diamagnetic (*i.e.* not attracted by a magnetic field).

Paramagnetism in the transition elements is caused by the presence of unpaired electrons in the d sub-orbital. Diamagnetism is characteristic of compounds where all the electrons are paired in the d sub-orbitals.

Properties of Transition Metals

Apart from Copper, the transition metals are all white lustrous metals. They vary widely in abundance (*e.g.* Iron, Fe, and Titanium, Ti, are plentiful, Scandium, Sc, is rare). They have high melting points and high densities. This suggests that the electrons which enter the d orbitals are being used to bind the atoms together in the crystal lattice. The transition elements form Complex Ions.

Examples of these compound ions include:

- Ferrocyanide Ion, Fe(CN)6(4–) in Potassium Ferrocyanide, K4Fe(CN)6,
- Chromate Ion, CrO4(2–), in Barium Chromate, BaCrO4.

Most compounds of the transition metal are coloured. There are variations in colour for compounds of the same valency, and with different valency (oxidation) states. For example,

A Coordination Complex is a compound in which molecules or ions form coordinate bonds to a central metal atom or ion. The complex may contain positive ions, negative ions or neutral molecules. The formation of such coordination complexes is typical behaviour of transition metals.

Diamagnetism is magnetic characteristic of those transition metals where all the electrons in the d sub-orbitals of the atoms are paired, and results in these elements not attracted by a magnetic field.

The Sub-Group Ib Elements are the transition metals Copper, Silver and Gold.

The Group III Elements in the periodic table are divided into two Sub-Groups. The Sub-Group IIIa Elements, which are also called the main group elements, are Boron, Aluminium, Gallium, Indium and Thallium.

The main group is essentially a group of weak metals, which have two s-electrons and one p-electron in their valence shells. Unlike the elements found in Groups I and in Group II, the Group III elements have more than one oxidation state, being the first members of the p block.

The Sub-Group IIIb Elements are the transition metals Scandium, Yttrium, Lanthanum, which is generally classified with the Lanthanoids and Actinium, which is generally classified with the Actinoids.

The Group IV Elements in the periodic table are divided into two Sub-Groups. The Sub-Group IVa Elements, which are also called the main group elements, are Carbon, Silicon, Germanium, Tin, and Lead.

These elements have two s electrons and two p electrons in their valence shells. The p electrons give rise to a +2 oxidation state while the s and p electrons give rise to a +4 oxidation state. The +4 oxidation state is formed due to an s electron being promoted to a vacant p orbital when the atom attains an excited state.

The Sub-Group IVb Elements consists of the three elements Titanium, Zirconium, and Hafnium. These three elements are generally classified with the transition elements. The Group V Elements in the periodic table are divided into two Sub-Groups.

The Sub-Group Va Elements, which are also called the main group elements, are Nitrogen, Phosphorus, Arsenic, Antimony, and Bismuth. There are five electrons in the valence shells of these elements (*i.e.* two s orbital electrons and three p orbital). Thus, the principal oxidation state of these elements are +3 and +5 for the main group elements.

The Sub-Group Vb Elements are Vanadium, Niobium, and Tantalum. These three elements are generally classified with the transition elements. The Group VI Elements in the periodic table are divided into two Sub-Groups. The Sub-Group VIa Elements, which are also called the main group elements, are Oxygen, Sulphur, Selenium, Tellurium, and Polonium. These elements have two electrons short of the electronic configuration of a Nobel Gas. Thus, the oxidation states +2, +4, and +6 are found in the elements of this group.

The Sub-Group VIb Elements are Chromium, Molybdenum and Tungsten. These three elements are generally classified with the transition elements.

The Group VII Elements in the periodic table are divided into two Sub-Groups. The Sub-Group VIIa Elements, which are also known as the Halogens (*i.e.* the main group elements), are Fluorine, Chlorine, Bromine, Iodine and Astatine. These elements are non-metals and are too reactive to occur on their own in nature. Thus, they are usually found combined with other elements in a salt. The Sub-Group VIIb Elements are the metallic transition elements Manganese, Technetium, and Rhenium. These three elements are generally classified with the transition elements.

The Group VIII Elements in the periodic table are a group of nine transition elements consisting of the platinum metals. This group consists of three triads of elements Iron, Ruthenium, Osmium, Cobalt, Rhodium, Iridium, and Nickel, Palladium Platinum. These elements are generally classified with the transition elements.

Inner Transition Metals

The Inner Transition Metals are the series of elements:

- From Cerium (Atomic Number 58) to Lutetium (Atomic Number 71), which are called the Lanthanoids,
- From Thorium (Atomic Number 90) to Lawrencium (Atomic Number 103), which are called the Actinoids.

The inner transition elements are found between Group 2 and the transition elements in the fifth row of the periodic table. The transition elements are also known as the f-block elements.

These two series make up the f-block elements in the periodic table, and their chemical properties of the elements derive from the filling of the f atomic sub-orbitals. The electronic configuration of these elements are characterised as having full outer orbitals and full second outermost orbitals, while the second outermost orbitals are incompletely filled. Thus, in the case of the first inner transition metals series, the electronic configuration of the outermost and second outermost orbitals is 4s2 3d10, while the third outermost orbitals (*i.e.* the 4f level) are incompletely filled.

The Main Group Elements in the periodic table can be classified as either being metals or non-metals. The vertical columns in the table are called groups. Elements within the same group all have the same number of electrons in their outer shell. Group 1 are known as Alkali metals. Group II are known as the Alkaline earth metals and the group of elements between group II and

III are known as the transition metals. Paramagnetism in the transition elements is caused by the presence of unpaired electrons in the d sub-orbital, and results in these elements being attracted by a magnetic field.

The colour exhibited by the transition metal ions is due to their ability to absorb light in the visible region of the electromagnetic spectrum. This causes unpaired d electrons to rise from a lower to a higher energy state. Ions which have the electronic configuration 3d(10), including the Scandium (III) Ion, Sc(+++) and the Copper (I) Ion, Cu(+), are colourless, because they have no unpaired d electrons.

Compound	*Formula*	*Colour*
Anhydrous Copper Sulphate	Cu(II)SO4,	White
Hydrated Copper Sulphate	Cu(II)SO4.5H2O	Light Blue
Ammoniacal Copper Sulphate	[Cu(II)(NH3)4]SO4	Dark Blue
Potassium Permanganate	KMnO4	Purple
Potassium Dichromate	KCrO4	Orange
Potassium Chromate	K2Cr2O4	Green
V(+++++) is pale yellow,		
V(++++) is blue,		

Compounds of the transition elements can be paramagnetic (*i.e.* be attracted by a magnetic field) or diamagnetic (*i.e.* not be attracted by a magnetic field). Paramagnetism in the transition elements is caused by the presence of unpaired electrons in the d orbitals. Diamagnetism is characteristic of compounds where all the electrons are paired in the d orbitals.

ATOMIC STRUCTURE: DISCOVERY OF ,Α Β & Γ PARTICLES

The discovery of x-rays by William Conrad Roentgen in November of 1895 excited the imagination of a generation of scientists who rushed to study this phenomenon. Within a few months, Henri Becquerel found that both uranium metal and salts of this element gave off a different form of radiation, which could also pass through solids.

By 1898, Marie Curie found that compounds of thorium were also "radioactive." After pain-staking effort she eventually isolated two more radioactive elements – polonium and radiumfrom ores that contained uranium. In 1899 Ernest Rutherford found that there were at least two different forms of radioactivity when he studied the absorption of radioactivity by thin sheets of metal foil. One, which he called alpha (α) particles, were absorbed by metal foil that was a few hundredths of a centimetre thick. The other, beta (β) particles, could pass through 100 times as much metal foil before they became absorbed. Shortly thereafter, a third form of radiation, gamma (γ) rays, was discovered that could penetrate as much as several centimetres of lead.

The results of early experiments on these three forms of radiation. The direction in which -particles were deflected by an electric field suggested that they were positively charged. The magnitude of this deflection suggested that they had the same charge-to-mass ratio as an He^{2+} ion.

To test the equivalence between -particles and He^{2+} ions, Rutherford built an apparatus that allowed -particles to pass through a very thin glass wall into an evacuated flask that contained a pair of metal electrodes. After a few days, he connected these electrodes to a battery and noted that the gas in the flask did indeed give off the characteristic emission spectrum of helium.

Experiments with electric and magnetic fields demonstrated that β particles were negatively charged. Furthermore, they had the same charge-to-mass ratio as an electron.

To date, no detectable difference has been found between β-particles and electrons. The only reason to retain the name " β-particle" is to emphasize the fact that these particles are ejected from the nucleus of an atom when it undergoes radioactive decay. The fact that -rays are not deflected by either electric or magnetic fields suggests that these rays don't carry an electric charge. Since they travel at the speed of light, they are classified as a form of electromagnetic radiation that carries even more energy than x-rays.

Properties of Radioactive Particles

Name	*Symbols*	*Identity*	*Relative Charge*	*Relative Mass*	*Penetrating Power*	*Interaction with Charged Plates*	*Hazards*
Alpha	α, $^{4}_{2}He$	Helium nucleus	2+	4	low - stopped by a sheet of paper	attracted to negative plate, deflected by positive plate	harmful if ingested
Beta	β, $^{0}_{-1}e$	Electron	1–	$\frac{1}{2000}$	moderate – pass through paper and ½ mm aluminium, stopped by ½ mm lead	attracted to positive plate, deflected by negative plate	skin burns, harmful if ingested particularly iodine-131 in thyroid & strontium-90 in bones
Gamma	γ	Electromagnetic radiation	0	0	high - only stopped by several centimetres of lead or many centimetres of concrete	unaffected	most dangerous as these are the most penetrating, as a consequence, gamma rays can be used to sterilize materials & destroy bacteria in food

Growth and Decay Curves

At the turn of the century, when radioactivity was discovered, atoms were assumed to be indestructible. Ernest Rutherford and Frederick Soddy, however, found that radioactive substances became less active with time. More importantly, they noticed that radioactivity was always accompanied by the formation of atoms of a different element. By 1903, they concluded that radioactivity was accompanied by a change in the structure of the atom. They therefore assumed that radiation was emitted when an element decayed into a different kind of atom.

By 1910, 40 radioactive elements had been isolated that were associated with the process by which uranium metal decayed to lead. This created a problem, however, because there was space for only 11 elements between lead and uranium. In 1913, Kasimir Fajans and Fredick Soddy proposed an explanation for these results based on the following rules.

- α -particles are emitted when an element is formed that belongs two spaces to the left in the periodic table. Uranium (Z = 92), for example, emits an α -particle when it decays to form thorium (Z = 90).
- β -particles are emitted when an element is formed that belongs one space to the right in the periodic table. Actinium (Z = 89) emits a β -particle when it decays to form thorium (Z = 90).
- Radioactive elements that fall in the same place in the periodic table are different forms of the same element. The radioactive thorium produced by the α-particle decay of uranium is a different form of the element than the radioactive thorium obtained by the β -particle decay of actinium.

Soddy proposed the name isotope to describe different radioactive atoms that occupy the same position in the periodic table. J. J. Thomson and Francis Aston then used a mass spectrometer to show that isotopes are atoms of the same element that have different atomic masses.

The Structure of the Atom

The discovery of the electron in 1897 by J. J. Thomson suggested that there was an internal structure to the "indivisible" building blocks of matter known as atoms. This raised an obvious question: How many electrons does an atom contain? By studying the scattering of light, x-rays, and -particles, Thomson concluded that the number of electrons in an atom was between 0.2 and 2 times the weight of the atom.

In 1911, Rutherford concluded that the scattering of -particles by extremely thin pieces of metal foil could be explained by assuming that all of the positive charge and most of the mass of the atom were concentrated in an infinitesimally small fraction of the total volume of the atom, for which he proposed the name nucleus.

Rutherford's data also suggested that the nucleus of a gold atom carries a positive charge that is about 80 times the charge on an electron.

The discovery of the neutron in 1932 explained the discrepancy between the charge on the nucleus and the mass of an atom. A neutral gold atom that has a mass of 197 amu consists of a nucleus that contains 79 protons and 118 neutrons surrounded by 79 electrons. By convention, this information is specified by the following symbol, which describes the only naturally occurring isotope of gold.

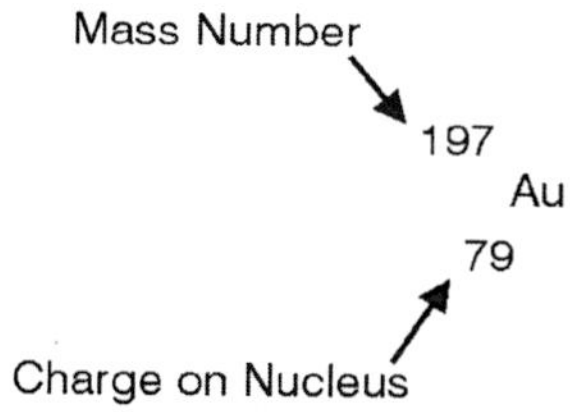

This convention can also be applied to subatomic particles. The only difference is the use of lowercase letters to identify the particle. Because anyone with access to a periodic table can find the atomic number of an element, a shorthand notation is often used that reports only the mass number of the atom and the symbol of the element. The shorthand notation for the naturally occurring isotope of gold is ^{197}Au.

NICOLSON: A CLEVER IDEA ABOUT A WRONG MODEL

Meanwhile, in Cambridge one J. W. Nicolson was struggling to incorporate Planck's ideas in a model of the atom, in an attempt to understand some strange sets of spectral lines observed in nebulae and in the sun's corona. He conceived a rather exotic model, in which a ring of electrons, like a necklace, orbited the nucleus.

The reason was that the classical radiation from a ring of electrons is a lot less that from a single orbiting electron, the fields tend to cancel each other. Oscillations of electrons in this ring gave the spectra. Nicolson predicted the frequencies emitted by a straightforward classical analysis of these oscillation frequencies, in the spirit of earlier work on the plum pudding model. He did bring in Planck's constant, though.

He knew that dimensionally it was a unit of angular momentum, and he suggested that the atom could only lose angular momentum in discrete amounts presumably constant multiples of h. Nicolson felt that, given the dimensionality of Planck's constant, quantization of angular momentum was more plausible than quantization of energy. Of course, for the simple harmonic oscillator they amounted to the same thing, but not for any other system.

Bohr Changes his Mind about Spectra

In 1913, Bohr was surprised to find out in a casual conversation with the spectroscopist H. R. Hansen that some patterns had been discerned in the apparent chaos of spectral lines. In particular, Hansen (a colleague and former classmate of Bohr) showed him Balmer's formula for hydrogen. They had very likely seen this in class together, but, given Bohr's opinion of the value of spectra, he probably hadn't paid much attention.

Hansen would doubtless have informed Bohr that the 1/4 could be replaced by $1/m^2$, with m another integer. The constant appearing on the right hand side is called the *Rydberg constant*, R_H = 109,737 cm^{-1}. (This is the modern value—Balmer got it right to one part in 10,000, about the limit of spectral measurements at the time.)

Bohr said later: "As soon as I saw Balmer's formula, the whole thing was immediately clear to me." What he saw was that the set of allowed *frequencies* proportional to inverse wavelengths emitted by the hydrogen atom could all be expressed as *differences*. This immediately suggested to him a generalization of his idea of a "stationary state" lowest energy level, in which the electron did not radiate. There must be a *whole sequence* of these stationary states, with radiation only taking place as the atom jumps from one to another of lower energy, emitting a single quantum of frequency f such that

$$hf = E_n - E_m,$$

the difference between the energies of the two states. Evidently, from the Balmer formula and its extension to general integers m, n, these allowed non-radiating orbits, the stationary states, could be labelled 1, 2, 3..., n... and had energies –1, –1/4, –1/9..., $-1/n^2$... in units of hcR_H. The energies are of course negative, because these are bound states, and we count energy zero from where the two particles are infinitely far apart.

Bohr was very familiar with the dynamics of simple circular orbits in an inverse square field. He knew that if the energy of the orbit was-hcR_H/n^2, that meant the kinetic energy of the electron, $\frac{1}{2}mv^2 = hcR_H/n^2$, and the potential energy would be

$$-(1/4\pi\varepsilon_0)e^2/r = -2hcR_H/n^2.$$

It immediately follows that the *radius* of the nth orbit is proportional to n^2, and the *speed* in that orbit is proportional to $1/n$. *It then follows that the angular momentum of the nth orbit is just proportional to n.*

Evidently, then the angular momentum in the nth orbit was nKh, where h is Planck's constant and K is some multiplying factor, the same for all the orbits, still to be determined. In fact, the value of K follows from the results above. R_H, m, h, and c are all known quantities (R_H being measured experimentally by observing the lines in the Balmer series) so the above formulas immediately give the electron's speed and distance from the nucleus in the nth orbit, and hence its angular momentum. Therefore, by putting in these experimentally determined quantities, we can find K.

Single electron in an atom

The most widely studied set of quantum numbers is that for a single electron in an atom: because it is not only useful in chemistry, being the basic notion behind the periodic table, valence and a host of other properties, but also because it is a solvable and realistic problem, and, as such, finds widespread. In non-relativistic quantum mechanics the Hamiltonian of this system consists of the kinetic energy of the electron and the potential energy due to the Coulomb force between the nucleus and the electron.

The kinetic energy can be separated into a piece which is due to angular momentum, J, of the electron around the nucleus, and the remainder. Since the potential is spherically symmetric, the full Hamiltonian commutes with J^2. J^2 itself commutes with any one of the components of the angular momentum vector, conventionally taken to be J_z. These are the only mutually commuting operators in this problem; hence, there are three quantum numbers.

These are conventionally known as:

- The principal quantum number (n = 1, 2, 3, 4...) denotes the eigenvalue of H with the J^2 part removed. This number therefore has a dependence only on the distance between the electron and the nucleus (ie, the radial coordinate, r). The average distance increases with n, and hence quantum states with different principal quantum numbers are said to belong to different shells.
- The azimuthal quantum number (l = 0, 1... n–1) (also known as the angular quantum number or orbital quantum number) gives the orbital angular momentum through the relation $L^2 = \hbar^2 l(l + 1)$. In chemistry, this quantum number is very important, since it specifies the shape of an atomic orbital and strongly influences chemical bonds and bond angles. In some contexts, l=0 is called an s orbital, l=1, a p orbital, l=2, a d orbital and l=3, an f orbital.
- The magnetic quantum number (m_l = $-l$, $-l$+1... 0... l–1, l) is the eigenvalue, $L_z = m_l\hbar$. This is the projection of the orbital angular momentum along a specified axis.

Results from spectroscopy indicated that up to two electrons can occupy a single orbital. However two electrons can never have the same exact quantum state nor the same set of quantum numbers according to Hund's Rules, which addresses the Pauli exclusion principle.

A fourth quantum number with two possible values was added as an *ad hoc* assumption to resolve the conflict; this supposition could later be explained in detail by relativistic quantum mechanics and from the results of the renown Stern-Gerlach experiment.

- The spin projection quantum number (m_s = –1/2 or +1/2), the intrinsic angular momentum of the electron. This is the projection of the spin s=1/2 along the specified axis.

Example: The quantum numbers used to refer to the outermost valence electron of the Fluorine (F) atom, which is located in the 2p atomic orbital, are; $n = 2$, $l = 1$, $m_l = 1$, or 0, or –1, $m_s = -1/2$ or 1/2. Note that molecular orbitals require totally different quantum numbers, because the Hamiltonian and its symmetries are quite different.

Quantum Numbers with Spin-orbit Interaction

When one takes the spin-orbit interaction into consideration, l, m and s no longer commute with the Hamiltonian, and their value therefore changes over time. Thus another set of quantum numbers should be used.

This set includes:

- The total angular momentum quantum number (j = 1/2,3/2... n–1/2) gives the total angular momentum through the relation $J^2 = \hbar^2 j(j + 1)$.
- The projection of the total angular momentum along a specified axis (m_j = -j,-j+1... j), which is analogous to m, and satisfies $m_j = m_l + m_s$.
- Parity. This is the eigenvalue under reflection, and is positive (*i.e.* +1) for states which came from even l and negative (*i.e.* -1) for states which came from odd l. The former is also known as even parity and the latter as odd parity

For example, consider the following eight states, defined by their quantum numbers:

- $l = 1$, $m_l = 1$, $m_s = +1/2$
- $l = 1$, $m_l = 1$, $m_s = -1/2$
- $l = 1$, $m_l = 0$, $m_s = +1/2$
- $l = 1$, $m_l = 0$, $m_s = -1/2$
- $l = 1$, $m_l = -1$, $m_s = +1/2$
- $l = 1$, $m_l = -1$, $m_s = -1/2$
- $l = 0$, $m_l = 0$, $m_s = +1/2$
- $l = 0$, $m_l = 0$, $m_s = -1/2$

The quantum states in the system can be described as linear combination of these eight states. However, in the presence of spin-orbit interaction, if one wants to describe the same system by eight states which are eigenvectors of the Hamiltonian, we should consider the following eight states:

- $j = 3/2$, $m_j = 3/2$, odd parity (coming from state (1))
- $j = 3/2$, $m_j = 1/2$, odd parity (coming from states (2) and (3))
- $j = 3/2$, $m_j = -1/2$, odd parity (coming from states (4) and (5))
- $j = 3/2$, $m_j = -3/2$, odd parity (coming from state (6))
- $j = 1/2$, $m_j = 1/2$, odd parity (coming from state (2) and (3))

- $j = 1/2$, $m_j = -1/2$, odd parity (coming from states (4r) and (5))
- $j = 1/2$, $m_j = 1/2$, even parity (coming from state (7))
- $j = 1/2$, $m_j = -1/2$, even parity (coming from state (8))

Elementary Particles

For a more complete description of the quantum states of elementary particles. Elementary particles contain many quantum numbers which are usually said to be intrinsic to them. However, it should be understood that the elementary particles are quantum states of the standard model of particle physics, and hence the quantum numbers of these particles bear the same relation to the Hamiltonian of this model as the quantum numbers of the Bohr atom does to its Hamiltonian.

In other words, each quantum number denotes a symmetry of the problem. It is more useful in field theory to distinguish between spacetime and internal symmetries. Typical quantum numbers related to spacetime symmetries are spin the parity, C-parity and T-parity.

Typical internal symmetries are lepton number and baryon number or the electric charge. It is worth mentioning here a minor but often confusing point. Most conserved quantum numbers are additive. Thus, in an elementary particle reaction, the sum of the quantum numbers should be the same before and after the reaction. However, some, usually called a *parity*, are multiplicative; ie, their product is conserved. All multiplicative quantum numbers belong to a symmetry in which applying the symmetry transformation twice is equivalent to doing nothing. These are all examples of an abstract group called Z_2.

ELECTRONIC STRUCTURES

								1 H									2 He
3 Li	4 Be											5 B	6 C	7 N	8 O	9 F	10 Ne
11 Na	12 Mg											13 Al	14 Si	15 P	16 S	17 Cl	18 Ar
19 K	20 Ca	21 Sc	22 Ti	23 V	24 Cr	25 Mn	26 Fe	27 Co	28 Ni	29 Cu	30 Zn	31 Ga	32 Ge	33 As	34 Se	35 Br	36 Kr

The First Period

Hydrogen has its only electron in the 1s orbital - $1s^1$, and at helium the first level is completely full - $1s^2$.

The Second Period

Now we need to start filling the second level, and hence start the second period. Lithium's electron goes into the 2s orbital because that has a lower energy than the 2p orbitals. Lithium has an electronic structure of $1s^2 2s^1$. Beryllium adds a second electron to this same level - $1s^2 2s^2$.

All the various p electrons can be lumped together. For example, fluorine could be written as $1s^2 2s^2 2p^5$, and neon as $1s^2 2s^2 2p^6$. This is what is normally done if the electrons are in an inner layer. If the electrons are in the bonding level (those on the outside of the atom), they are sometimes written in shorthand, sometimes in full. Don't worry about this. Be prepared to meet either version, but if you are asked for the electronic structure of something in an exam, write it out in full showing all the p_x, p_y and p_z orbitals in the outer level separately. For example, although we haven't yet met the electronic structure of chlorine, you could write it as $1s^2 2s^2 2p^6 3s^2 3p_x^2 3p_y^2 3p_z^1$.

The 2p electrons are all lumped together whereas the 3p ones are shown in full. The logic is that the 3p electrons will be involved in bonding because they are on the outside of the atom, whereas the 2p electrons are buried deep in the atom and aren't really of any interest.

All the inner electrons together using, for example, the symbol [Ne]. In this context, [Ne] means *the electronic structure of neon* - in other words: $1s^2 2s^2 2p_x^2 2p_y^2 2p_z^2$ You wouldn't do this with helium because it takes longer to write [He] than it does $1s^2$. On this basis the structure of chlorine would be written $[Ne]3s^2 3p_x^2 3p_y^2 3p_z^1$.

The Third Period

At neon, all the second level orbitals are full, and so after this we have to start the third period with sodium. The pattern of filling is now exactly the same as in the previous period, except that everything is now happening at the 3-level.

For example:

		short version
Mg	$1s^2 2s^2 2p^6 3s^2$	$[Ne]3s^2$
S	$1s^2 2s^2 2p^6 3s^2 3p_x^2 3p_y^1 3p_z^1$	$[Ne]3s^2 3p_x^2 3p_y^1 3p_z^1$
Ar	$1s^2 2s^2 2p^6 3s^2 3p_x^2 3p_y^2 3p_z^2$	$[Ne]3s^2 3p_x^2 3p_y^2 3p_z^2$

The Beginning of the Fourth Period

At this point the 3-level orbitals aren't all full - the 3d levels haven't been used yet. But if you refer back to the energies of the orbitals, you will see that the next lowest energy orbital is the 4s - so that fills next.

K $1s^2 2s^2 2p^6 3s^2 3p^6 4s^1$ Ca

$1s^2 2s^2 2p^6 3s^2 3p^6 4s^2$

There is strong evidence for this in the similarities in the chemistry of elements like sodium ($1s^2 2s^2 2p^6 3s^1$) and potassium ($1s^2 2s^2 2p^6 3s^2 3p^6 4s^1$)

The outer electron governs their properties and that electron is in the same sort of orbital in both of the elements. That wouldn't be true if the outer electron in potassium was $3d^1$.

The elements in group 1 of the Periodic Table all have an outer electronic structure of ns^1 (where n is a number between 2 and 7). All group 2 elements have an outer electronic structure of ns^2. Elements in groups 1 and 2 are described as s-block elements. Elements from group 3 across to the noble gases all have their outer electrons in p orbitals. These are then described as p-block elements.

D-BLOCK ELEMENTS

Remember that the 4s orbital has a lower energy than the 3d orbitals and so fills first. Once the 3d orbitals have filled up, the next electrons go into the 4p orbitals as you would expected-block elements are elements in which the last electron to be added to the atom is in a d orbital. The first series of these contains the elements from scandium to zinc, which at GCSE you probably called transition elements or transition metals. The terms "transition element" and "d-block element" don't quite have the same meaning, but it doesn't matter in the present context. d electrons are almost always described as, for example, d^5 or d^8 - and not written as separate orbitals. Remember that there are five d orbitals, and that the electrons will inhabit them singly as far as possible. Up to 5 electrons will occupy orbitals on their own. After that they will have to pair up.

d^5 means

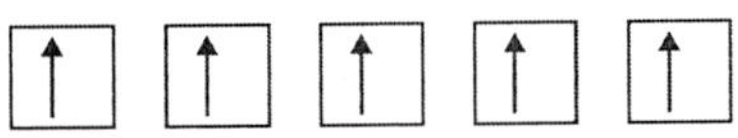

d^8 means

Notice in what follows that all the 3-level orbitals are written together, even though the 3d electrons are added to the atom after the 4s.

Sc $1s^22s^22p^63s^23p^63d^14s^2$

Ti $1s^22s^22p^63s^23p^63d^24s^2$

V $1s^22s^22p^63s^23p^63d^34s^2$

Cr $1s^22s^22p^63s^23p^63d^54s^1$

Whoops! Chromium breaks the sequence. In chromium, the electrons in the 3d and 4s orbitals rearrange so that there is one electron in each orbital. It would be convenient if the sequence was tidy - but it's not!

Mn $1s^22s^22p^63s^23p^63d^54s^2$ (back to being tidy again)

Fe $1s^22s^22p^63s^23p^63d^64s^2$

Co $1s^2 2s^2 2p^6 3s^2 3p^6 3d^7 4s^2$

Ni $1s^2 2s^2 2p^6 3s^2 3p^6 3d^8 4s^2$

Cu $1s^2 2s^2 2p^6 3s^2 3p^6 3d^{10} 4s^1$

(another awkward one!)

Zn $1s^2 2s^2 2p^6 3s^2 3p^6 3d^{10} 4s^2$

And at zinc the process of filling the d orbitals is complete.

Atomic Orbitals

Atomic orbitals are in a way that makes them understandable for introductory courses and its equivalents. It explores s and p orbitals in some detail, including their shapes and energies. d orbitals are described only in terms of their energy, and f orbitals only get a passing mention.

Orbitals and Orbits

When the a planet moves around the sun, you can plot a definite path for it which is called an orbit. A simple view of the atom looks similar and you may have pictured the electrons as orbiting around the nucleus. The truth is different, and electrons in fact inhabit regions of space known as *orbitals.* Orbits and orbitals sound similar, but they have quite different meanings. It is essential that you understand the difference between them.

The *Heisenberg Uncertainty Principle* says - loosely - that you can't know with certainty both where an electron is and where it's going next. It is impossible to define with absolute precision, at the same time, both the position and the momentum of an electron.

That makes it impossible to plot an orbit for an electron around a nucleus. Is this a big problem? No. If something is impossible, you have to accept it and find a way around it.

Hydrogen's Electron - the 1s Orbital

Suppose you had a single hydrogen atom and at a particular instant plotted the position of the one electron. Soon afterwards, you do the same thing, and find that it is in a new position. You have no idea how it got from the first place to the second.In the hydrogen case, the electron can be found anywhere within a spherical space surrounding the nucleus. The diagram shows a *cross-section* through this spherical space.95% of the time (or any other percentage you choose), the electron will be found within a fairly easily defined region of space quite close to the nucleus. Such a region of space is called an *orbital.*

You can think of an orbital as being the region of space in which the electron lives. What is the electron doing in the orbital? We don't know, we can't know, and so we just ignore the problem! All you can say is that if an electron is in a particular orbital it will have a particular definable energy. Each orbital has a name. The orbital occupied by the hydrogen electron is called a *1s orbital.* The *"1"* represents the fact that the orbital is in the energy level closest to the nucleus. The *"s"* tells you about the shape of the orbital. s orbitals are spherically symmetric around the nucleus - in each case, like a hollow ball made of rather chunky material with the nucleus at its centre.

The orbital on the left is a *2s orbital.* This is similar to a 1s orbital except that the region where there is the greatest chance of finding the electron is further from the nucleus - this is an orbital at the second energy level.

If you look carefully, you will notice that there is another region of slightly higher electron density (where the dots are thicker) nearer the nucleus. ("Electron density" is another way of talking about how likely you are to find an electron at a particular place.) 2s (and 3s, 4s, etc.) electrons spend some of their time closer to the nucleus than you might expect.

The effect of this is to slightly reduce the energy of electrons in s orbitals. The nearer the nucleus the electrons get, the lower their energy. 3s, 4s (etc.) orbitals get progressively further from the nucleus.

p Orbitals

Not all electrons inhabit s orbitals (in fact, very few electrons live in s orbitals). At the first energy level, the only orbital available to electrons is the 1s orbital, but at the second level, as well as a 2s orbital, there are also orbitals called *2p orbitals. A* p orbital is rather like 2 identical balloons tied together at the nucleus. The diagram on the right is a cross-section through that 3-dimensional region of space. Once again, the orbital shows where there is a 95% chance of finding a particular electron.

Unlike an s orbital, a p orbital points in a particular direction. At any one energy level it is possible to have three absolutely equivalent p orbitals pointing mutually at right angles to each other.

These are arbitrarily given the symbols p_x, p_y and p_z. This is simply for convenience - what you might think of as the x, y or z direction changes constantly as the atom tumbles in space.

The p orbitals at the second energy level are called $2p_x$, $2p_y$ and $2p_z$. There are similar orbitals at subsequent levels - $3p_x$, $3p_y$, $3p_z$, $4p_x$, $4p_y$, $4p_z$ and so on. All levels except for the first level have p orbitals. At the higher levels the lobes get more elongated, with the most likely place to find the electron more distant from the nucleus.

d and f Orbitals

In addition to s and p orbitals, there are two other sets of orbitals which become available for electrons to inhabit at higher energy levels.

At the third level, there is a set of five d orbitals (with complicated shapes and names) as well as the 3s and 3p orbitals ($3p_x$, $3p_y$, $3p_z$). At the third level there are a total of nine orbitals altogether. At the fourth level, as well the 4s and 4p and 4d orbitals there are an additional seven f orbitals - 16 orbitals in all. s, p, d and f orbitals are then available at all higher energy levels as well.

For the moment, you need to be aware that there are sets of five d orbitals at levels from the third level upwards, but you probably won't be expected to draw them or name them. Apart from a passing reference, you won't come across f orbitals at all.

Fitting Electrons into Orbitals

You can think of an atom as a very bizarre house (like an inverted pyramid!) - with the nucleus living on the ground floor, and then various rooms (orbitals) on the higher floors occupied by the electrons. On the first floor there is only 1 room (the 1s orbital); on the second floor there are 4 rooms (the 2s, $2p_x$, $2p_y$ and $2p_z$ orbitals); on the third floor there are 9 rooms (one 3s orbital, three 3p orbitals and five 3d orbitals); and so on. But the rooms aren't very big... Each orbital can only hold 2 electrons. A convenient way of showing the orbitals that the electrons live in is to draw "electrons-in-boxes".

"Electrons-in-Boxes"

Orbitals can be represented as boxes with the electrons in them shown as arrows. Often an up-arrow and a down-arrow are used to show that the electrons are in some way different.

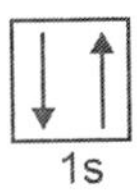

A 1s orbital holding 2 electrons would be drawn as shown on the right, but it can be written even more quickly as $1s^2$. This is read as "one s two" - not as "one s squared". You mustn't confuse the two numbers in this notation:

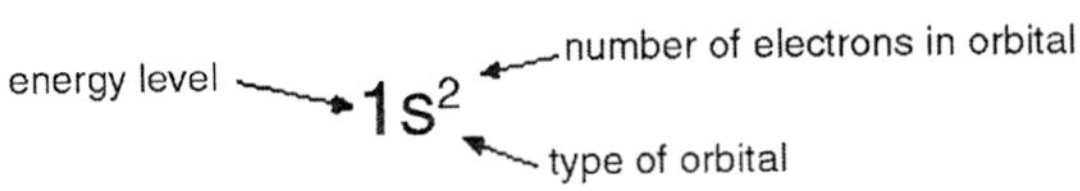

The Order of Filling Orbitals

Electrons fill low energy orbitals (closer to the nucleus) before they fill higher energy ones. Where there is a choice between orbitals of equal energy, they fill the orbitals singly as far as possible. This filling of orbitals singly where possible is known as *Hund's rule*. It only applies where the orbitals have exactly the same energies (as with p orbitals, for example), and helps to minimise the repulsions between electrons and so makes the atom more stable.

RADIOACTIVE DECAY

α β and Radiation

Early studies of radioactivity indicated that three different kinds of radiation were emitted, symbolized by the first three letters of the Greek alphabet, β and. With time, it became apparent that this classification scheme was much too simple. The emission of a negatively charged β particle, for example, is only one example of a family of radioactive transformations known as β -decay. A fourth category, known as spontaneous fission, also had to be added to describe the process by which certain radioactive nuclides decompose into fragments of different weight.

Alpha Decay

*A*lpha decay is usually restricted to the heavier elements in the periodic table. (Only a handful of nuclides with atomic numbers less than 83 emit an -particle.) The product of -decay is easy to predict if we assume that both mass and charge are conserved in nuclear reactions. Alpha decay of the ^{238}U "parent" nuclide, for example, produces ^{234}Th as the "daughter" nuclide.

$$\begin{matrix} 238 & & 234 & & 4 & & \\ & U \rightarrow & & Th + & & He & (\alpha\text{-}decay) \\ 92 & & 90 & & 2 & & \end{matrix}$$

The sum of the mass numbers of the products (234 + 4) is equal to the mass number of the parent nuclide (238), and the sum of the charges on the products (90 + 2) is equal to the charge on the parent nuclide.

Beta Decay

There are three different modes of beta decay:

- electron (b^-) emission
- electron capture
- positron (b^+) emission

Electron (β^-) emission is literally the process in which an electron is ejected or emitted from the nucleus. When this happens, the charge on the nucleus increases by one. Electron (β^-) emitters are found throughout the periodic table, from the lightest elements (^3H) to the heaviest (^{255}Es). The product of β^--emission can be predicted by assuming that both mass number and charge are conserved in nuclear reactions. If ^{40}K is a β^--emitter, for example, the product of this reaction must be ^{40}Ca.

$$^{40}_{19}K \rightarrow ^{40}_{20}Ca + ^{0}_{-1}e \qquad [electron(\beta^-)emission]$$

Once again the sum of the mass numbers of the products is equal to the mass number of the parent nuclide and the sum of the charge on the products is equal to the charge on the parent nuclide. Nuclei can also decay by capturing one of the electrons that surround the nucleus. Electron capture leads to a decrease of one in the charge on the nucleus.

The energy given off in this reaction is carried by an x-ray photon, which is represented by the symbol hv, where h is Planck's constant and v is the frequency of the x-ray.

The product of this reaction can be predicted, once again, by assuming that mass and charge are conserved.

$$^{40}_{19}K + ^{0}_{20}e \rightarrow ^{40}_{-1}Ar + h\nu \qquad (electron\ capture)$$

The electron captured by the nucleus in this reaction is usually a 1s electron because electrons in this orbital are the closest to the nucleus. A third form of beta decay is called positron (β^+) emission.

The positron is the antimatter equivalent of an electron. It has the same mass as an electron, but the opposite charge. Positron (β^+) decay produces a daughter nuclide with one less positive charge on the nucleus than the parent.

$$^{40}_{19}K \rightarrow ^{40}_{18}Ar + ^{0}_{+1}e \qquad [positron(\beta^+)emission]$$

Positrons have a very short life-time. They rapidly lose their kinetic energy as they pass through matter. As soon as they come to rest, they combine with an electron to form two γ-ray photons in a matter-antimatter annihilation reaction.

$$^{0}_{+1}e + ^{0}_{-1}e \rightarrow 2\gamma$$

Thus, although it is theoretically possible to observe a fourth mode of beta decay corresponding to the capture of a positron, this reaction does not occur in nature. Note that in all three forms of β -decay for the ^{40}K nuclide the mass number of the parent and daughter nuclides are the same for electron emission, electron capture, and position emission. All three forms of β -decay therefore interconvert isobars.

Gamma Emission

The daughter nuclides produced by α-decay or β-decay are often obtained in an excited state. The excess energy associated with this excited state is released when the nucleus emits a photon in the γ-ray portion of the electromagnetic spectrum.

Most of the time, the γ-ray is emitted within 10^{-12} seconds after the α-particle or β-particle.

In some cases, gamma decay is delayed, and a short-lived, or metastable, nuclide is formed, which is identified by a small letter m written after the mass number. ^{60m}Co, for example, is produced by the electron emission of ^{60}Fe.

$$^{60}_{26}Fe \rightarrow \, ^{60m}_{27}Co + \, ^{0}_{-1}e$$

The metastable ^{60m}Co nuclide has a half-life of 10.5 minutes. Since electromagnetic radiation carries neither charge nor mass, the product of γ-ray emission by ^{60m}Co is ^{60}Co.

$$^{60m}_{27}Co \rightarrow \, ^{60}_{27}Co + \gamma \quad (\gamma\text{-}ray\, emission)$$

Spontaneous Fission

Nuclides with atomic numbers of 90 or more undergo a form of radioactive decay known as spontaneous fission in which the parent nucleus splits into a pair of smaller nuclei. The reaction is usually accompanied by the ejection of one or more neutrons.

$$^{252}_{98}Cf \rightarrow \, ^{140}_{54}Xe + \, ^{108}_{44}Ru + 4\, ^{1}_{0}n$$

For all but the very heaviest isotopes, spontaneous fission is a very slow reaction. Spontaneous fission of ^{238}U, for example, is almost two million times slower than the rate at which this nuclide undergoes α-decay.

Neutro n-Rich Versus Neutron-Poor Nuclides

In 1934 Enrico Fermi proposed a theory that explained the three forms of beta decay. He argued that a neutron could decay to form a proton by emitting an electron. A proton, on the other hand, could be transformed into a neutron by two pathways. It can capture an electron or it can emit a positron.

Electron emission therefore leads to an increase in the atomic number of the nucleus.

$$^{14}_{6}C \rightarrow \, ^{14}_{7}N + \, ^{0}_{-1}e$$

Both electron capture and positron emission, on the other hand, result in a decrease in the atomic number of the nucleus.

$$^{7}_{4}Be + \, ^{0}_{-1}e \rightarrow \, ^{7}_{3}Li + h\nu$$

$$^{11}_{6}C \rightarrow \, ^{11}_{5}B + \, ^{0}_{+1}e$$

A plot of the number of neutrons versus the number of protons for all of the stable naturally occurring isotopes. Several conclusions can be drawn from this plot.

- The stable nuclides lie in a very narrow band of neutron-to-proton ratios.
- The ratio of neutrons to protons in stable nuclides gradually increases as the number of protons in the nucleus increases.
- Light nuclides, such as ^{12}C, contain about the same number of neutrons and protons. Heavy nuclides, such as ^{238}U, contain up to 1.6 times as many neutrons as protons.
- There are no stable nuclides with atomic numbers larger than 83.
- This narrow band of stable nuclei is surrounded by a sea of instability.
- Nuclei that lie above this line have too many neutrons and are therefore neutron-rich.
- Nuclei that lie below this line don't have enough neutrons and are therefore neutron-poor.

The most likely mode of decay for a neutron-rich nucleus is one that converts a neutron into a proton. Every neutron-rich radioactive isotope with an atomic number smaller 83 decays by electron (β^-) emission. ^{14}C, ^{32}P, and ^{35}S, for example, are all neutron-rich nuclei that decay by the emission of an electron.

$$^{35}_{16}S \rightarrow \, ^{35}_{17}Cl + \, ^{0}_{-1}e \quad (\beta^- \textit{ emission})$$

Neutron-poor nuclides decay by modes that convert a proton into a neutron. Neutron-poor nuclides with atomic numbers less than 83 tend to decay by either electron capture or positron emission. Many of these nuclides decay by both routes, but positron emission is more often observed in the lighter nuclides, such as ^{22}Na.

$$^{22}_{11}Na \rightarrow \, ^{22}_{10}Ne + \, ^{0}_{+1}e \quad (\beta^+ \; emission)$$

Electron capture is more common among heavier nuclides, such as ^{125}I, because the 1s electrons are held closer to the nucleus of an atom as the charge on the nucleus increases.

$$^{125}_{53}I + \, ^{0}_{-1}e \rightarrow \, ^{125}_{52}Te + h\nu \quad (electron\,capture)$$

A third mode of decay is observed in neutron-poor nuclides that have atomic numbers larger than 83. Although it is not obvious at first, -decay increases the ratio of neutrons to protons. Consider what happens during the -decay of ^{238}U, for example.

$$^{238}_{92}U \rightarrow \, ^{234}_{90}Th + \, ^{4}_{2}He \quad (\alpha\, decay)$$

The parent nuclide (^{238}U) in this reaction has 92 protons and 146 neutrons, which means that the neutron-to-proton ratio is 1.587. The daughter nuclide (^{234}Th) has 90 protons and 144 neutrons, so its neutron-to-proton ratio is 1.600. The daughter nuclide is therefore slightly less likely to be neutron-poor.

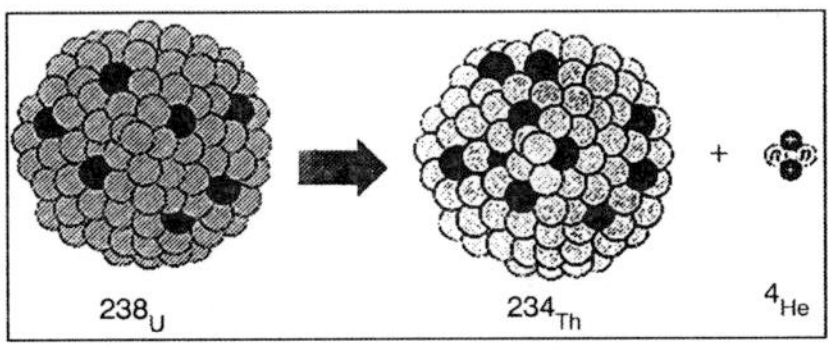

$$\left(\frac{146\,\text{neutrons}}{92\ \text{protons}} = 1.587\right) \qquad \left(\frac{144\ \text{neutrons}}{90\ \text{protons}} = 1.600\right)$$

Binding Energy Calculations

We should be able to predict the mass of an atom from the masses of the subatomic particles it contains. A helium atom, for example, contains two protons, two neutrons, and two electrons.

Example:

The mass of a helium atom should be 4.0329802 amu.

2(1.0072765) amu = 2.0145530 amu

2(1.0086650) amu = 2.0173300 amu

2(0.0005486) amu = 0.0010972 amu

Total mass = 4.0329802 amu

When the mass of a helium atom is measured, we find that the experimental value is smaller than the predicted mass by 0.0303769 amu.

Predicted mass = 4.0329802 amu

Observed mass = 4.0026033 amu

Mass defect = 0.0303769 amu

The difference between the mass of an atom and the sum of the masses of its protons, neutrons, and electrons is called the mass defect. The mass defect of an atom reflects the stability of the nucleus. It is equal to the energy released when the nucleus is formed from its protons and neutrons. The mass defect is therefore also known as the binding energy of the nucleus.

The binding energy serves the same function for nuclear reactions as H for a chemical reaction. It measures the difference between the stability of the products of the reaction and the starting materials. The larger the binding energy, the more stable the nucleus. The binding energy can also be viewed as the amount of energy it would take to rip the nucleus apart to form isolated neutrons and protons. It is therefore literally the energy that binds together the neutrons and protons in the nucleus. The binding energy of a nuclide can be calculated from its mass defect with Einstein's equation that relates mass and energy.

$E = mc^2$

Example: We found the mass defect of He to be 0.0303769 amu. To obtain the binding energy in units of joules, we must convert the mass defect from atomic mass units to kilograms.

$$0.030769\,amu\,x\frac{1.6605655x10^{-24}\,g}{1\,amu}x\frac{1kg}{1000g} = 5.04428x10^{-29}\,kg$$

Multiplying the mass defect in kilograms by the square of the speed of light in units of meters per second gives a binding energy for a single helium atom of 4.53358 x 10^{-12} joules.

$$E = \left(5.04428x10^{-29}\,kg\right)\left(2.9979246x10^{8}\,m/s\right)^2 = 4.53358x10^{-12}\,J$$

Multiplying the result of this calculation by the number of atoms in a mole gives a binding energy for helium of 2.730 x 10^{12} joules per mole, or 2.730 billion kilojoules per mole.

$$\frac{4.5358x10^{-12}\,J}{1\,atom}x\frac{6.022x10^{23}\,atoms}{1\,mol} = 2.730x10^{12}\,Jimol$$

This calculation helps us understand the fascination of nuclear reactions. The energy released when natural gas is burned is about 800 kJ/mol. The synthesis of a mole of helium releases 3.4 million times as much energy. Since most nuclear reactions are carried out on very small samples of material, the mole is not a reasonable basis of measurement. Binding energies are usually expressed in units of electron volts (eV) or million electron volts (MeV) per atom.

Example: The binding energy of helium is 28.3×10^6 eV/atom or 28.3 MeV/ atom.

$$\frac{4.5358x10^{-12}J}{1atom}x\frac{1ev}{1.6021892x10^{-19}J} = 28.30x10^6 evlatom$$

Calculations of the binding energy can be simplified by using the following conversion factor between the mass defect in atomic mass units and the binding energy in million electron volts.

1 amu = 931.5016 MeV

Binding energies gradually increase with atomic number, although they tend to level off near the end of the periodic table. A more useful quantity is obtained by dividing the binding energy for a nuclide by the total number of protons and neutrons it contains. This quantity is known as the binding energy per nucleon. The binding energy per nucleon ranges from about 7.5 to 8.8 MeV for most nuclei.

It reaches a maximum, however, at an atomic mass of about 60 amu. The largest binding energy per nucleon is observed for ^{56}Fe, which is the most stable nuclide in the periodic table.

The graph of binding energy per nucleon versus atomic mass explains why energy is released when relatively small nuclei combine to form larger nuclei in fusion reactions.

$$^{12}_{6}C + ^{12}_{6}C \rightarrow ^{24}_{12}Mg \quad (fusion)$$

It also explains why energy is released when relatively heavy nuclei split apart in fission (literally, "to split or cleave") reactions.

$$^{235}_{92}U \rightarrow ^{139}_{56}Ba + ^{94}_{36}Kr + 2\,^{1}_{0}n \quad (fission)$$

There are a number of small irregularities in the binding energy curve at the low end of the mass spectrum. The ^{4}He nucleus, for example, is much more stable than its nearest neighbors. The unusual stability of the ^{4}He nucleus explains why -particle decay is usually much faster than the spontaneous fission of a nuclide into two large fragments.

THE KINETICS OF RADIOACTIVE DECAY

Radioactive nuclei decay by first-order kinetics. The rate of radioactive decay is therefore the product of a rate constant (k) times the number of atoms of the isotope in the sample (N).

Rate = kN

The rate of radioactive decay doesn't depend on the chemical state of the isotope. The rate of decay of ^{238}U, for example, is exactly the same in uranium metal and uranium hexafluoride, or any other compound of this element. The rate at which a radioactive isotope decays is called the activity of the isotope.

The most common unit of activity is the curie (Ci), which was originally defined as the number of disintegrations per second in 1 gram of ^{226}Ra. The curie is now defined as the amount of radioactive isotope necessary to achieve an activity of 3.700×10^{10} disintegrations per second.

The relative rates at which radioactive nuclei decay can be expressed in terms of either the rate constants for the decay or the half-lives of the nuclei. We can conclude that ^{14}C decays more rapidly than ^{238}U, for example, by noting that the rate constant for the decay of ^{14}C is much larger than that for ^{238}U.

^{14}C:		$k = 1.210 \times 10^{-4}\ y^{-1}$
^{238}U:		$k = 1.54 \times 10^{-10}\ y^{-1}$

We can reach the same conclusion by noting that the half-life for the decay of ^{14}C is much shorter than that for ^{235}U.

^{14}C:		$t_{1/2} = 5730$ y
^{238}U:		$t_{1/2} = 4.51 \times 10^{9}$ y

The half-life for the decay of a radioactive nuclide is the length of time it takes for exactly half of the nuclei in the sample to decay. In our discussion of the kinetics of chemical reactions, we concluded that the half-life of a first-order process is inversely proportional to the rate constant for this process.

$$t_{1/2} = \frac{\text{In } 2}{\text{k}} = \frac{0.693}{k}$$

The half-life of a nuclide can be used to estimate the amount of a radioactive isotope left after a given number of half-lives.

Dating By Radioactive Decay

The earth is constantly `bombarded by cosmic rays emitted by the sun. The total energy received in the form of cosmic rays is small no more than the energy received by the planet from starlight. But the energy of a single cosmic ray is very large, on the order of several billion electron volts (0?200 million kJ/mol). These highly energetic rays react with atoms in the atmosphere to produce neutrons that then react with nitrogen atoms in the atmosphere to produce ^{14}C.

$$^{14}_{7}N + {}^{1}_{0}n \rightarrow {}^{14}_{6}C + {}^{1}_{1}H$$

The ^{14}C formed in this reaction is a neutron-rich nuclide that decays by electron emission with a half-life of 5730 years.

$$^{14}_{6}C \rightarrow {}^{14}_{7}N + {}^{0}_{-1}e$$

Just after World War II, Willard F. Libby proposed a way to use these reactions to estimate the age of carbon-containing substances. The ^{14}C dating technique for which Libby received the Nobel prize was based on the following assumptions.

- ^{14}C is produced in the atmosphere at a more or less constant rate.
- Carbon atoms circulate between the atmosphere, the oceans, and living organisms at a rate very much faster than they decay. As a result, there is a constant concentration of ^{14}C in all living things.
- After death, organisms no longer pick up ^{14}C.

Thus, by comparing the activity of a sample with the activity of living tissue we can estimate how long it has been since the organism died.

The natural abundance of ^{14}C is about 1 part in 10^{12} and the average activity of living tissue is 15.3 disintegrations per minute per gram of carbon. Samples used for ^{14}C dating can include charcoal, wood, cloth, paper, sea shells, limestone, flesh, hair, soil, peat, and bone. Since most iron samples also contain carbon, it is possible to estimate the time since iron was last fired by analysing for ^{14}C.

We now know that one of Libby's assumptions is questionable: The amount of ^{14}C in the atmosphere hasn't been constant with time. Because of changes in solar activity and the earth's magnetic field, it has varied by as much as 5 per cent. More recently, contamination from the burning of fossil fuels and the testing of nuclear weapons has caused significant changes in the amount of radioactive carbon in the atmosphere. Radiocarbon dates are therefore reported in years before the present era (B.P.). By convention, the present era is assumed to begin in 1950, when ^{14}C dating was introduced.

Studies of bristlecone pines allow us to correct for changes in the abundance of ^{14}C with time. These remarkable trees, which grow in the White Mountains of California, can live for up to five thousand years. By studying the ^{14}C

activity of samples taken from the annual growth rings in these trees, researchers have developed a calibration curve for ^{14}C dates from the present back to 5145 B.C.

After roughly 45,000 years (eight half-lives), a sample retains only 0.4 per cent of the ^{14}C activity of living tissue. At that point it becomes too old to date by radiocarbon techniques. Other radioactive isotopes can be used to date rocks, soils, or archaeological objects that are much older. Potassium-argon dating, for example, has been used to date samples up to 4.3 billion years old. Naturally occurring potassium contains 0.0118 per cent by weight of the radioactive ^{40}K isotope.

This isotope decays to ^{40}Ar with a half-life of 1.3 billion years. The ^{40}Ar produced after a rock crystallizes is trapped in the crystal lattice. It can be released, however, when the rock is melted at temperatures up to 2000C. By measuring the amount of ^{40}Ar released when the rock is melted and comparing it with the amount of potassium in the sample, the time since the rock crystallized can be determined.

Atom Symbols

An atom of Helium that had 2 protons, 2 neutrons, and 2 electrons. Its called Helium 4. Now we will see the other symbols that are used to identify this kind of atom. The diagram to the left or above shows two common types of symbols.

The Top symbol is just a shortened version of Helium 4. Most of the time the chemical names are not spelled out. Instead, the chemical symbol (He) is used. The number that follows the name or the symbol is the mass number. The mass number tells us which Helium isotope we are looking at by telling how many protons and neutrons are in the nucleus.

If we look at a periodic table of the elements we see that Helium always has an atomic number of 2, which tells us that there are two protons in this atom. So we can subtract the Atomic number (2) from the mass number (4) to get 2 neutrons in He 4. The other symbol (on the graphic) already shows both the Mass number and the Atomic number. By the way, the Atomic Mass is not the same as the Mass Number. The Atomic Mass is an average mass of naturally occurring isotopes that are found on the earth. If you have a sample of Helium, the average mass would be 4.0026. Mass Number, on the other hand, is the protons + neutrons of an atom. So do not confuse the two terms.

Isotopes of Helium

There are many isotopes of Helium. To the left or above, we can see the two stable isotopes that were already introduced above. In the Laboratory, scientists are able to make other isotopes (or varieties) of Helium but these are often unstable and they decay into something else. This makes them radioactive.

These are extremely short lived isotopes (varieties). Helium 5 has 3 neutrons and has a half-life of 0.0000000000000000000002 seconds. So Helium 5 does not stick around very long. The half-life is the measured prediction of the time it takes for half of the atoms in a sample to decay and turn into something else in a radioactive process. Helium 6 has 4 neutrons and is somewhat more stable than Helium 5. Half of Helium 6 will decay in 0.81 seconds.

How many protons are in He 8 is easy, we just use the atomic number, 2 protons. Many neutrons are in He 8, we would use the following equation: Mass number-atomic number = neutrons (8–2 = 6).

Atomic Size

A big problem with the nuclear hydrogen atom was: what determined its size?

Now this equation is satisfied by *any* circular orbit centred at the nucleus, however large or small. Multiplying both sides by $r/2$ gives that the magnitude of the kinetic energy in the circular orbit is just half the magnitude of the negative potential energy.

There is no hint here that the atom in its "natural" ground state should have any particular radius. Bohr (and others) thought that Planck's constant must somehow play a role in determining the size of the orbit. After all, it *did* play a role in restricting allowed orbital changes in the oscillators in black body radiation and these oscillators, although not very clearly understood, were of the same general size as atoms.

So evidently the standard picture of how an oscillating charge radiated couldn't be right at the atomic level. Bohr concluded that in an atom in its natural rest state, the electron must be in a special orbit, he called it a "stationary state" to which the usual rules of electromagnetic radiation didn't apply. In this orbit, which determined the size of the atom, the electron, mysteriously, didn't radiate.

Just how to bring Planck's constant into a discussion of the hydrogen atom was not so clear, though. For the black body oscillators, it related the frequency f of the oscillator with the allowed energy change E by $E = hf$.

The obvious parallel approach for the hydrogen atom was to identify the frequency f with the circular frequency of the electron in its orbit. However, in contrast to the simple harmonic oscillator this hydrogen atom frequency *varied* with the size of the orbit. Still, it was the only frequency around, and, dimensionally, multiplying it by h gave an energy. What energy could that be identified with? Again, the choice was limited—the electron had a kinetic energy E, the potential energy was $-2E$ and the total energy $-E$. If a hydrogen nucleus captured a passing electron into its ground state, and emitted one quantum of electromagnetic radiation, that quantum would have energy E, the same as the electron kinetic energy in the natural stationary state (called the *ground state*).

Bohr suggested in a note to Rutherford in 1912 that requiring this energy be some constant assumed to be of order of magnitude one multiplied by hf would fix the size of the atom. Actually his argument was a bit more complicated, he considered the several electron atom, and took the electrons to form rings. However, the basic point is the same a condition like this constrains the atomic size, it would be fixed uniquely if we knew the constant.

Putting this together with the dynamic equation above determines the atomic radius r. It is easy to check that it predicts a radius of $4\pi\varepsilon_0 h^2/\pi^2 me^2$, which is just four times the "right answer" defined as the Bohr radiusse. The correct Bohr radius comes out if we choose the constant to be one-half, $E = \frac{1}{2}hf$, which Bohr used later.

Hence the approximate size of the atom follows from *dimensional* arguments alone once one assumes that Planck's constant plays a role! Of course, the nucleus is irrelevant in determining the atomic size—it just provides a fixed centre of electrostatic attraction.

The relevant electronic parameters are the mass m and the strength of attraction $e^2/4\pi\varepsilon_0$. Together with h, these parameters determine a length. It should be mentioned that this assumption explained more than the size of the hydrogen atom. It was believed at the time that in the higher atoms, the electrons formed rings, thought to lie one outside the other, and various stability arguments indicated that there couldn't be more than seven electrons in a ring.

The length scale above,

$$4\pi\varepsilon_0 h^2/\pi^2 me^2,$$

would *decrease* for larger atoms, with e^2 replaced by Ze^2 essentially, for nuclear charge Z. Thus as the number of rings increased, the size of the rings would decrease, explaining the observed approximate periodicity in atomic volume with atomic number. Also, in 1911 Richard Whiddington in Cambridge had found that to cause a substance having atomic number A to emit characteristic x-rays by bombarding it with electrons, it was necessary to use electrons of speed approximately $A \times 10^6$ meters per second. Any substance on being bombarded with sufficiently fast electrons emits a continuum of x-ray frequencies up to a maximum frequency f given by hf = kinetic energy of electron, *plus* some sharply defined lines—x-rays at a particular frequency, which does not change as the electron speed is further increased.

The frequency corresponding to these lines was found to increase with atomic number. Applying his length scale argument to the innermost ring of an atom, Bohr found that an electron in that ring would have a speed proportional to the nuclear charge, and hence, at least approximately, to the atomic number. Furthermore, the predicted speed in orbit was of the same order as that of Whiddington's electrons.

3

Chemical Bonding

Chemical bonds are the attractive forces that hold atoms together in the form of compounds. They are formed when electrons are shared between two atoms. There are 3 types of bonds...covalent bonds, polar covalent bonds and ionic bonds. The simplest example of bonding can be demonstrated by the H_2 molecule.

We can see from the periodic table that each hydrogen atom has a single electron. If 2 hydrogen atoms come together to form a bond, then each hydrogen atom effectively has a share in both electrons and thus each resembles an inert gas and is more stabile. The 2 electrons that are shared can be represented either by 2 dots or a single dash between the atoms.

$$H\cdot + \cdot H \rightarrow H:H \text{ or } H-H$$

Valence bond theory describes a chemical bond as the overlap of atomic orbitals. In the case of the hydrogen molecule, the 1s orbital of one hydrogen atom overlaps with the 1s orbital of the second hydrogen atom to form a molecular orbital called a sigma bond.

Attraction increases as the distance between the atoms gets closer but nuclear-nuclear repulsion becomes important if the atoms approach too close.

There are 3 methods of showing the formulas of molecules. Molecular formulas show only the types and numbers of atoms in the molecule. Structural formulas show the atoms in their correct placement in the molecule and allow for distinguishing isomers. Electron-dot formulas are similar to structural formulas but also include all of the non-bonding outer electrons. Knowledge of electron placement allows us to understand not only the shape of molecules but their chemical character. If we understand the chemical character of a molecule, we can predict how it will react with other molecules without having to blindly memorize reactions.

CHEMICAL BONDS

Before Dalton's atomic theory there wasn't much reason to consider how atoms stick together, although it was clear that some force is necessary to bind compounds of elements together. The term "chemical affinity" referred more to the ability to react (undergo a chemical change) than to form a stable compound. When it was realised that chemical compounds were formed of discrete molecules containing fixed numbers of atoms, the need to explain molecular stability developed, leading to the notion of the interatomic chemical bond (Latin inter for among plus band for a fastening) holding atoms together.

Early Nineteenth Century speculations about the origin of interatomic bonds were affected by contemporary discoveries in electricity, indicating that "opposites" attract. Positive and negative charges are electrical opposites, and north and south poles are magnetic opposites. Precisely just what constituted chemical opposites wasn't clear, however. Chemists had classified substances into metals and non-metals, and into acids and bases, both of which seemed opposite in the sense that they had "affinity" for each other and readily reacted together to form new compounds.

Yet there was little evidence to support the hypothesis of Avogadro that non-metal gases were composed of two atoms of the same element, nor were there any explanations for the solid state of metallic elements, or the extreme hardness of non-metallic elements such as diamond. The number of bonds an element could make, called the valence (Latin valere for strength) of the element, is suggested by the atom ratios of chemical formulas, and ordinarily limited to a small number. For example, hydrogen atoms never forms bonds to more than one another atom, so are assigned a valency of one.

Oxygen has a valency of two (is divalent) as each oxygen atom combines with two hydrogen atoms in forming water (H_2O, more descriptively written showing atom connections as H-O-H). Carbon has a valency of four (is tetravalent) because it combines with a maximum of four hydrogen atoms (in methane, CH4 more descriptively shown with C surrounded by 4 H), or, equivalently, a maximum of two oxygen atoms, each with a valency of two (in carbon dioxide, CO_2, more descriptively written O-C-O).

The electrons of atoms are electromagnetically attracted by the nuclei of atoms, due to the opposite electric charge of electrons and nuclei. Chemical bonds are characterised by physical states in which a few electrons move partly from one atom to one or more other atoms, driven by the achievement of a lower state of energy from this motion.

This lowering of energy is caused by a rearrangement of charges, usually resulting in net decrease in the average distance between the electrons of all the bonded atoms, and their nuclei. The transfer of charge caused by the movement of the electron from one atom to another, also causes the participating atoms (which may number from two to many) to be attracted to one another electromagnetically. The attractive force between atoms is the bond. Chemical bonds, for the sake of simplicity, are classically assigned characteristics of two major types: covalent and ionic.

In a simplified view of a pure covalent type bond, the bond forms as a few electrons farthest from their atomic nuclei become more attracted to the region of space between two nuclei. In this region, negatively-charged electrons experience attraction from the positively-charged protons from more than one nucleus. This causes some electrons to spend a high probability of their time in the interatomic space. In turn, the nuclei are stabilized in position by the pull from these shared electrons during the fraction of time that the bonding electrons reside between the atoms.

Although such bonding electrons do not spend all of their time between atoms, when they spend more time between a given pair of atoms than otherwise, they constitute chemical bonds. Nuclei fixed by such bonds may vibrate, but they are pulled towards each other by the mutual forces of the bonding electrons pulling them together, yet prevented from approaching too closely by their own charge, or else by the mutual repulsion of other inner electrons, which are held so closely and tightly to individual nuclei that they cannot be shared to any important degree.

In a simplified view of a pure ionic type bond, one or more outer electrons are not shared between atoms, but donated from one atom to another.

In such a bond, the structure of the electron cloud of one of the nuclei contains an available space for another electron, which allows an additional electron to experience a greater net attraction from the nucleus than is experienced by outer electrons in a neighbouring atom towards their own nucleus.

This difference in available states causes effective transfer of one or more electrons from one atom to another atom, where they can be more tightly bound. This transfer causes the donating atom to assume a net positive charge, and the other to assume a net negative charge; the atoms thus become positive or negatively charged ions. The bond then results from electrostatic attraction between these ionized atoms.

Most bonds have a mixture of covalent and ionic character, as bonding electrons are shared between atoms, but shared somewhat unevenly. All bonds can be explained by quantum theory, but in practice, simplification rules allow chemists to predict the strength, directionality, and polarity of bonds. The octet rule and VSEPR theory are two examples. More sophisticated theories are valence bond theory which includes orbital hybridization and resonance, and the linear combination of atomic orbitals molecular orbital method which includes ligand field theory. Electrostatics are used to describe bond polarities and the effects they have on chemical substances.

COVALENT BONDS

Don't look now, but you're surrounded by covalent bonds. They're in the air you breathe and in the water you drink. You even make them yourself when you exhale. What are these bonds? What do they want?

Take a look at water (H_2O). H_2O is a Molecule, a discrete unit of atoms that are bonded together. A molecule of water consists of an oxygen atom that is bonded to two hydrogen atoms. A hydrogen atom and the oxygen atom each donate one valence electron to form a chemical bond. These two valence electrons forming the bond are shared by both atoms, resulting in a Single Covalent Bond. Think of this in terms of two pieces of wood that are nailed together. The pieces of wood are the atoms, and the nails holding them together are the electrons that form the covalent bond. Each piece of wood shares a portion of the nails.

Air, too, contains oxygen. Oxygen does not exist as a single oxygen atom, but as a molecule of two oxygen atoms. These two oxygen atoms share two pairs of valence electrons (four valence electrons total) between them, forming a Double Covalent Bond. This is true of any double covalent bond; four valence electrons are shared between two atoms.

Another component of air is nitrogen. Like oxygen, nitrogen does not exist as a single nitrogen atom, but as a molecule made up of two nitrogen atoms. The two nitrogen atoms in a molecule of nitrogen share three pairs of valence electrons (six valence electrons total) to form a Triple Covalent Bond. Is it possible to predict whether bonds are covalent or not? A good rule of thumb is that bonds between nonmetals (remember that hydrogen is considered a nonmetal) are usually covalent bonds. For example, the carbon dioxide (CO_2) molecules you exhale are bonded together covalently.

Fatal Attraction: Ionic Bonds

Now that you've gotten over being surrounded by covalent bonds, it's time you knew something else. You ate ionic bonds. Lots. Today. You can find some in your saltshaker. In salt (sodium chloride), atoms do not share electrons. Rather, one type of atom strips at least one valence electron from another type of atom, creating ions of opposite charges. These two oppositely charged atoms are held together by an Electrostatic Interaction, an attraction between oppositely charged particles. An Ionic Bond is an electrostatic interaction that holds together a positively charged ion (cation) and a negatively charged ion (anion). In an ionic bond, one atom loses an electron to another atom, forming a cation and anion, respectively. And, as everyone knows, opposites attract.

In table salt, for example, a valence electron from a sodium atom is transferred to a chlorine atom, forming Na+ and Cl-. Because the ions have opposite charges, they are attracted to each other. The loss of a valence electron and the attraction to the atom that took it happen simultaneously. It is possible for more than one valence electron to be drawn away from another atom, as in barium chloride ($BaCl_2$, a substance used in medicinal preparations). In barium chloride, two chlorine atoms each take one valence electron away from barium, leaving the barium ion Ba^{2+}.

Unlike water or oxygen, which have covalent bonds, substances that have ionic bonds do not exist naturally as discrete molecules. Rather, they form Ionic Solids, three-dimensional networks in which each cation is surrounded by anions and each anion is surrounded by cations. Look again in the saltshaker. Each sodium cation is surrounded or ionically bonded to six chloride anions, and each chloride anion is ionically bonded to six sodium cations. The formula NaCl for sodium chloride shows that for every sodium atom present in a piece of salt, there is one chlorine atom present. What, don't see any ionic bonds? All right, it looks like this:

Where are ionic bonds found? These types of bonds usually form when metal atoms bond with nonmetal atoms. In salt, the metal sodium bonds with the nonmetal chlorine. Besides salt, some other examples are lithium fluoride (LiF), strontium oxide (SrO) and calcium chloride ($CaCl_2$).

Ules of the Atomic Playground: Polar Covalent Bonds

Atoms, like preschoolers, don't always play fair. This means that sometimes in a covalent bond the electrons are not shared equally between the two atoms. On average, one of the atoms partially "pulls" the bonding electrons toward itself, creating an unequal sharing of those bonding electrons. This is called a Polar Covalent Bond. In order to determine whether a covalent bond is polar or not, it's necessary to understand electronegativity.

Recall that electronegativity is a measure of an atom's ability to draw its bonding electrons to itself. Each element has a numeric value corresponding to its electronegativity. The values used here were devised by Linus Pauling, though there are a few other scales of electronegativity values.

Fluorine was determined to be the most electronegative element and has an electronegativity value of 4.0. Francium, the least electronegative element, has a value of 0.7. It is important to remember a general trend in the periodic table: electronegativity increases from left to right going across a period, and it increases from the bottom to the top of a group. For example, in period 2, nitrogen (group 5A) has an electronegativity value of 3.0, compared to 2.5 for carbon (group 4A). In group 5A, phosphorous (period 3) has an electronegativity value of 2.1, compared to 3.0 for nitrogen (period 2).

In an action that resembles toddlers tugging on a toy, a polar covalent bond occurs when one atom with a higher electronegativity draws the bonding electrons toward itself, pulling those electrons away from the atom with the lower electronegativity value. This creates an unequal sharing of electrons known as Unequal Charge Distribution, or charge separation. The charge separation makes the bond polar because the more electronegative atom becomes partially negatively charged and the atom with the lower electronegativity becomes partially positively charged. Consider H_2 and HBr. In H_2, the atoms have an equal "pull" on the bonding electrons, making the bond Nonpolar. In HBr, however, Br has an electronegativity of 2.8, compared to 2.1 for hydrogen. The Br atom pulls the bonding electrons toward itself, creating a partial negative charge on itself and a partial positive charge on the hydrogen atom. The larger the absolute difference in electronegativity between two atoms, the more polar that bond is. For example, the electronegativity difference of a carbon-oxygen bond is -1.0, the result of 2.5 (the electronegativity value for carbon) minus 3.5 (the value for oxygen). The absolute value for the difference in electronegativity is the value without the minus sign (1.0 for a carbon-oxygen bond). For a carbon-chlorine bond, the difference in electronegativity is 0.5 (2.5-3.0 = -0.5). Thus, a carbon-oxygen bond is more polar (1.0) than a carbon-chlorine bond (0.5).

PROPERTIES OF SP-HYBRID ORBITALS

(i) Two sp-hybrid orbitals [sp_a and sp_b] are Completely equivalent in energy.

(ii) Each sp hybrid orbital is stronger than the pure s– and pure p-orbital from which it is formed after hybridisation. Its predicated relative overlapping power is 1.93.

(iii) In sp-hybrid orbital, there is one large lobe and one small lobe. The larger lobe, which is much bigger than the p-orbital, brings about a higher degree of overlapping with the orbital of a reacting atom. Thus, the resultant bond is comparatively stronger.

(iv) The normalised wave functions, ψ (sp_a) and for sp_b hybrid orbital are as follows :

$$\psi\,(sp_a) = \frac{1}{\sqrt{(2)}}\,[\psi\,(s) + \psi\,(p)]$$

$$\psi\,(sp_b) = \frac{1}{\sqrt{(2)}}\,[\psi\,(s) - \psi\,(p)]$$

From the above equations, it follows that the ratio of p-character to s-character is 1 : 1 in each of the two hybrid orbital wave functions and that both the hybrid orbital wave functions are orthogonal to each other.

Example of sp-hybridisation : An interesting example involving sp-hybridisation is beryllium fluoride, BeF_2.

LDQR THEORY

In 1961, Linnett proposed an interesting modification of Lewis-Langmuir octed theory. According to him,

(i) The eight valency electrons present around atoms of first short period in their molecules should not be treated as four pairs but should be treated as two groups of four electrons. Each group of four electrons is termed as *quartet.* As eight electrons are assumed to be made up of two quartets in this theory, it is named as *Linnett double quartet theory (LDQT).*

(ii) Linnett further proposed that a set of four electrons (quartet) be considered to have the same spin, □, another quartet to have the opposite spin, ↓. In a quartet, the four electrons will naturally tend to arrange themselves in a way that their mutual repulsion effects are minimised. This is possible in the tetrahedral configuration.
The second quartet will also take up the tetrahedral configuration. Thus, the eight electrons arrange themselves in two tetrahedra. The two tetrahedra take up relative positions that keep the inter-electronic repulsion at a minimum.

(iii) This theory also considers the resonance contributions. However, it works best when an equal number of single and double bonds are present (SO_2 or CO_2).

(iv) This theory reduces the emphasis on electron pairing.

The mathematical treatment of Linnett theory is yet in the formative stage. This theory is intermediate between the VBT and MOT.

Let us illustrate the Linnett theory by considering the following examples:

Nitrogen Molecule : There are ten valency electrons between two nitrogen atoms in their molecule. This molecule can be represented by the Lewis and Linnett structures as follows :

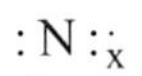

Lewis

Linnett

According to both Lewis and Linnett the molecule is diamagnetic (paired spin) and the bond order is 3, *i.e.*, (6 – 0), The LDQ representation where in there are two sets of five paired electrons.

The structure of carbon monoxide, acetylide ion (C_2^{2-}) and (CN^-) are similar.

Fluorine Molecule : In fluorine molecule, there are fourteen valency electrons. The Lewis structure of fluorine molecule is as follows :

```
        xx
: N̈:ₓ  F x
        xx
```

Lewis structure

In the LDQ structure, each fluorine atom has a quartet of electrons having the same spin, *i.e.*, four electrons denoted by x and four others by o. All the fourteen electrons are arranged in two spin sets of seven electrons. As the spins are equal in number and therefore cancel one another (7↑ and 7↑), the molecule is diamagnetic. Again, there are two electrons between the fluorine nuclei and hence bond order is one.

Oxygen Molecule : In this molecule there are sixteen electrons. According to the Lewis structure this molecule should be diamagnetic but actually it is paramagnetic.

:Ö::Ö:

Lewis structure

Linnett proposed the following structure :

```
x       x       x
.       .       .
x   O   x   O   x
.       .       .
x       x       x
.       .       .
```

The above structure of Linnett actually represents two states for oxygen molecule. These are :

(a) One is the high energy state, ¹Sg⁺.. In this structure the two twin states are completely identical.

(b) The excited state. Its energy state lies 130 kJl mol^{-1} above the ground state. When this loses energy, it becomes stabilized giving die staggered arrangement ($^1\Delta g$) which is also diamagnetic because the electrons are still paired 6↑, 6↑.

(c) If the structure is formed from one spin set of five electrons and the second of seven (5↑, 7↑) the molecules is further stabilized. This is the $^3\Sigma g^-$ state and is more stable because its energy is at the ground state. In this state the repulsion has been minimised, in spite of the total bonding electrons being four in number. In this structure there are two unpaired electrons and this shows that this molecule is paramagnetic.

Nitric Oxide Molecule : In this molecule there are 11 valency electrons having a set of 5 electrons of one spin (I) and 6 electrons of the opposite spin set (II). Each set has a tetrahedral orientation around the two atoms just as in O_2.

· : O. ˣₓNˣₓ O ˣₓ

(II)

The two structure (I) and (II) combine to yield structure (III).

· ˣₓNˣₓ : Oˣₓ· or ˣN ≐ O ˣ

(III)

The structure (III) accounts for the presence of one unpaired electron in NO. The observed bond length in NO is intermediate between that expected for a double bond and a triple bond.

This is in agreement with the bond order of 2 1/2 in the LDQ representation of NO (structure III).

Now the question arises why NO does not exist as $(NO)_2$ because the nitric oxide molecule (structure III) has an odd number of electrons. But dimeric form of this molecule is not possible due to the following reasons.

(i) From the thermodynamic point of view, entropy is in favour of small molecules,

(ii) Dimerisation would take place if there is some gain in bonding. In the dimerisation of NO there is no gain of electrons because the bond order of 2 1/2 *i.e.* 1/2 (5–0) remains the same, in the dimer. If the dimer $(NO)_2$ were to be formed, one might expect it to dimerise through the pairing of odd electrons:

O = N = N = O =

It is valid that the dimer is unstable relative to the monomer.

EXPERIMENTAL EVIDENCE IN SUPPORT OF WERNER'S THEORY

A large number of complexes were studied by Werner and other workers, Werner used molar conductance values to determine the number of ions per molecule thus assigning groups to the first co-ordination sphere. Chemical methods like precipitation of Cl^- by $AgNO_3$ to determine the number of ionisable CF ions per molecule were also used.

Cobalt (III) ammine complexes : When an aqueous cobalt (II) chloride solution is treated with ammonia and then oxidised by air, the following six compounds are isolated :

(a) Luteo cobaltic chloride, $CoCl_3{\cdot}_6NH_3$: This is an orange yellow crystalline compounds. Its chief characteristics are :

(i) *Treatment of this compound in the solid state with sulphuric acid liberates all the chlorines as hydrogen chloride and leaves a sulphate.* $Co_2(SO_4)_3.12NH_3$. *Treatment of its solution with silver nitrate precipitates all the chlorines immediately. Thus, the bonding between cobalt and chlorine is purely ionic.*

(ii) *Treatment of solid with hydrochloric acid even at 100°C affects no removal of ammonia. Similarly, treatment of the solid with moist silver oxide gives a water soluble compound of composition* $CoOOH.6NH_3$ *which on addition of acids reforms the salts of type*

From the above experimental facts it follows that all the six ammonia molecules are unionisable and are held to the metal ion Co^{3+} in the first or inner coordination sphere whereas the ionisable Cl^- ions are in outer coordination sphere. From the above considerations, Werner formulated this complex as:

$[Co(NH_3)_6]Cl_3$

Conductance measurement indicates the presence of four ions, *i.e.*, one cobalt and three chloride ions (ammonia molecules are neutral). This is in accordance with the dissociation,

$$[Co(NH_3)_6\ Cl_3 \rightleftharpoons [Co(NH_3)_6]^{3+} + 3Cl^-$$

(b) Roseo cobaltic Chloride $CoCl_3.5NH_3.H_2O$: This is a pink crystalline compound with the following characteristics :

(i) With $AgNO_3$, all the chlorines are precipitated as chloride from the solution of this complex indicating their ionizable character.

(ii) The water present in compound is firmly held at room temperature and is lost only at 100°C or above, showing its presence in the inner coordination sphere.

(iii) On treating with conc. H_2SO_4 it yields a compound $Co_2(SO_4)_3\ 10NH_3.5H_2O$ which on slight heating loses three water molecules, forming $Co_2(SO_4)_3.10NH_3.2H_2O$.

The above facts clearly indicate that water molecules and five ammonia molecules are firmly bound with cobalt whereas chlorines are loosely bound with cobalt. Thus, the formula of complex is

$[Co(NH_3)_5.H_2O]Cl_3$

Conductance measurements indicate the presence of four ions per molecule in very dilute solutions. This is in accord with the dissociation.

$$[Co(NH_3)_5.H_2O]Cl_3 \rightleftharpoons [Co(NH_3)5.H_2O]^{3+} + 3Cl^-$$

(c) Purpureo cobaltic chloride $CoCl_3.5NH_3$: It is a violet coloured compound with the following characteristics:

(i) When treated with conc. H_2SO_4, two chlorine atoms are lost as HCl, forming $Co(Cl)SO_4$.

(ii) Only two chlorines are precipitated immediately as AgCl on adding silver nitrate to its solution.

The above mentioned facts indicate the one chlorine atom is different from the rest two chlorine atoms. It means that only two of the three chlorines are ionisable as Cl^- ions.

On the basis of the above facts, Werner postulated the following formula : $[Co(NH_3)_5CL]Cl_2$

Conductance measurements indicate the presence of three ions per molecule in very dilute solutions. This is in accord with the dissociation.

$$[Co(NH_3)_5Cl]Cl_2 \rightleftharpoons [Co(NH_3)_5Cl]^{2+} + 2Cl^-$$

(d and e) Praseo-cobaltic chloride, violeo cobaltic chloride. $CoCl_3.4NH_3$: Two related compounds, one violet in colour (violeo-cobaltic chloride) and the other green (praseo-cobaltic chloride), have been prepared. Both have the same composition $CoCl_3.4NH_3$ but the chemical behaviour of these two compounds is quite different.

The conductances of solutions of these compounds show the presence of only two ions and from them only one chlorine is immediately precipitated by silver nitrate. Thus, from structural point of view their composition may be written as, $[Co(NH_3)_4Cl_2]Cl$

However, there were distinct differences in chemical behaviour between the two. Thus, although, only one third chlorine gets precipitated immediately on the addition of silver nitrate, second and third choleric atoms per molecule get also precipitated slowly and after long standing and no boiling.

Werner explained this by assuming that the chlorines of inner co-ordination sphere are slowly replaced by water molecules :

$$[Co(NH_3)_4Cl_2]^+ + H_2O \longrightarrow [Co(NH_3)_4H_2O.Cl]^{2+} + Cl^-$$

$$[Co(NH_3)_4H_2O.Cl]^{2+} + H_2O \longrightarrow [Co(NH_3)_4.2H_2O]^{3+} + Cl^-$$

However, the rate of precipitation of these chlorines in these compounds is different.

Werner explained the above facts by assuming that praseo-cobaltic chloride is trans- type whereas violeo-cobaltic chloride is cis- type.

(f) $CoCl_3.3NH_3$: This is a blue green compound having the following characteristics :

(i) *It does not give any precipitate immediately on addition of silver nitrate to its solution (absence of ionisable chlorine).*

(ii) *Its dilute solutions did not conduct electricity at all, showing the absence of any ions in solution.*

From the above facts Werner came to the important conclusion that all the groups are linked to the central metal ion by unionisable secondary valencies and gave the following formula : $[Co(NH_3)_3Cl_3]$

Werner applied his theory to the above ammonia compounds of cobalt (III) chloride. In all the compounds discussed above, the primary valencies are designated by solid lines and secondary valencies by dotted lines. In each case, satisfaction of six secondary valencies is essential. Werner's theory is successful in

(i) Predicting the correct number of ions in each case.

(ii) Firm association of chlorine with cobalt in the last three cases and

(iii) The existence of two forms of $CoCl_3.4NH_3$ is explained. It is due to difference in three dimensional arrangement of four ammonias and two chlorines about the central cobalt (III) ion.

In order to differentiate, materials held by secondary valencies are enclosed in square brackets when writing formulae whereas materials held by primary valencies outside the brackets. Thus the compounds just considered are written as.

$[Co(NH_3)_6]Cl_3)$, $[Co(NH_3)_5(H_2O)]\ Cl_3$, $[Co(NH_3)_5Cl]Cl_2$ $[Co(NH_3)_4Cl_2]\ Cl$ and $[Co(NH_3)_3Cl_3]$

Criticism of Werner's Theory : One must appreciate the remarkable work done by Werner. He did this work 70 years ago and inspite of advances in theory and the increase in the number of coordination compounds, his postulates are as valid today as when they were presented over 70 years ago.

BAND STRUCTURE OF ELECTRONIC

The electronic band structure (or simply band structure) of a solid describes ranges of energy that an electron is "forbidden" or "allowed" to have. It is due to the diffraction of the quantum mechanical electron waves in the periodic crystal lattice. The band structure of a material determines several characteristics, in particular its electronic and optical properties.

Bands occur in materials

The electrons of a single free-standing atom occupy atomic orbitals, which form a discrete set of energy levels. If several atoms are brought together into a molecule, their atomic orbitals split, as in a coupled oscillation. This produces a number of molecular orbitals proportional to the number of atoms. When a large number of atoms (of order 1020 or more) are brought together to form a solid, the number of orbitals becomes exceedingly large, and the difference in energy between them becomes very small, so the levels may be considered to form continuous bands of energy rather than the discrete energy levels of the atoms in isolation. However, some intervals of energy contain no orbitals, no matter how many atoms are aggregated, forming band gaps.

Within an energy band, energy levels are so numerous as to be a near continuum. First, the separation between energy levels in a solid is comparable with the energy that electrons constantly exchange with phonons (atomic vibrations). Second, it is comparable with the energy uncertainty due to the Heisenberg uncertainty principle, for reasonably long intervals of time. As a result, the separation between energy levels is of no consequence.

Basic Concepts

Any solid has a large number of bands. In theory, it can be said to have infinitely many bands. However, all but a few lie at energies so high that any electron that reaches those energies escapes from the solid. These bands are usually disregarded. Bands have different widths, based upon the properties of the atomic orbitals from which they arise. Also, allowed bands may overlap, producing (for practical purposes) a single large band. Metals contain a band that is partly empty and partly filled regardless of temperature. Therefore they have very high conductivity. The lowermost, almost fully occupied band in an insulator or semiconductor is called the valence band by analogy with the valence electrons of individual atoms. The uppermost, almost unoccupied band is called the conduction band because only when electrons are excited to the conduction band can current flow in these materials.

The difference between insulators and semiconductors is only that the forbidden band gap between the valence band and conduction band is larger in an insulator, so that fewer electrons are found there and the electrical conductivity is lower. Because one of the main mechanisms for electrons to be excited to the conduction band is due to thermal energy, the conductivity of semiconductors is strongly dependent on the temperature of the material. This band gap is one of the most useful aspects of the band structure, as it strongly influences the electrical and optical properties of the material. Electrons can transfer from one band to the other by means of carrier generation and recombination processes. The band gap and defect states created in the band gap by doping can be used to create semiconductor devices such as solar cells, diodes, transistors, laser diodes, and others.

Symmetry

A more complete view of the band structure takes into account the periodic nature of a crystal lattice using the symmetry operations that form a space group. The Schrφdinger equation is solved for the crystal, which has Bloch waves as solutions:

$$\psi_{nk}(r) = e^{ik.r}u_{nk}(r),$$

where *k* is called the wavevector, and is related to the direction of motion of the electron in the crystal, and n is the band index, which simply numbers the energy bands. The wavevector k takes on values within the Brillouin zone (BZ) corresponding to the crystal lattice, and particular directions/points in the BZ are assigned conventional names like Γ, Δ, Λ, Σ, etc. These directions are shown for the face-centred cubic lattice geometry.

The available energies for the electron also depend upon k, for silicon in the more complex energy band diagram at the right. In this diagram the topmost energy of the valence band is labeled Ev and the bottom energy in the conduction band is labeled Ec.

The top of the valence band is not directly below the bottom of the conduction band (Ev is for an electron travelling in direction Γ, Ec in direction X), so silicon is called an indirect gap material. For an electron to be excited from the valence band to the conduction band, it needs something to give it energy Ec "Ev and a change in direction/momentum. In other semiconductors (for example GaAs) both are at Γ, and these materials are called direct gap materials (no momentum change required). Direct gap materials benefit the operation of semiconductor laser diodes. Anderson's rule is used to align band diagrams between two different semiconductors in contact.

Band Structures in Different Types of Solids

Although electronic band structures are usually associated with crystalline materials, quasi-crystalline and amorphous solids may also exhibit band structures. However, the periodic nature and symmetrical properties of crystalline materials makes it much easier to examine the band structures of these materials theoretically. In addition, the well-defined symmetry axes of crystalline materials makes it possible to determine the dispersion relationship between the momentum (a 3-dimension vector quantity) and energy of a material. As a result, virtually all of the existing theoretical work on the electronic band structure of solids has focused on crystalline materials.

Density of States

While the density of energy states in a band could be very large for some materials, it may not be uniform. It approaches zero at the band boundaries, and is generally highest near the middle of a band. The density of states for the free electron model in three dimensions is given by,

$$D(\in) = \frac{V}{2\pi^2}\left(\frac{2m}{\hbar^2}\right)^{3/2} \in^{1/2}$$

Filling of Bands

Although the number of states in all of the bands is effectively infinite, in an uncharged material the number of electrons is equal only to the number of protons in the atoms of the material. Therefore not all of the states are occupied by electrons ("filled") at any time. The likelihood of any particular state being filled at any temperature is given by the Fermi-Dirac statistics.

The probability is given by the following:

$$f(E) = \frac{1}{1 + e^{\frac{E - E_F}{k_B T}}}$$

where:

- k_B is Boltzmann's constant,
- T is the temperature,
- E_F is the Fermi energy (or 'Fermi level').

The Fermi level naturally is the level at which the electrons and protons are balanced.

At T = 0, the distribution is a simple step function:

$$f(E) = \begin{cases} 1 & \text{if } 0 < E \leq E_F \\ 0 & \text{if } E_F < E \end{cases}$$

At nonzero temperatures, the step "smooths out", so that an appreciable number of states below the Fermi level are empty, and some states above the Fermi level are filled.

DEPICTING CHEMICAL STRUCTURES

Chemical species are traditionally represented by structural formulas such as the one for ascorbic acid (vitamin C) which we show here. The lines, of course, represent the "chemical bonds" of the molecule. More importantly, the structural formula of a molecule defines its connectivity, as was illustrated in the comparison of ethanol and dimethyl ether shown above. One limitation of such formulas is that they are drawn on a two-dimensional paper or screen, whereas most molecules have a three-dimensional shape. The wedge-shaped lines in the structural formula are one way of indicating which bonds extend above or below the viewing plane. You will probably be spared having to learn this convention until you get into second-year Chemistry.

Three-dimensional models (either real plastic ones or images that incorporate perspective and shading) reveal much more about a molecule's structure. The ball-and-stick and space-filling renditions are widely employed, but each has its limitations, as seen in the following examples:

The image produced by the AFM probe is shown at the very bottom. What is actually being imaged is the surface of the electron clouds of the molecule, which consists of six fused hexagonal rings of carbon atoms with hydrogens on its periphery. The tiny bumps that correspond to these hydrogen atom attest to the remarkable resolution of this experiment.

Visualization of Molecular Structures

The purpose of rendering a molecular structure in a particular way is not to achieve "realism" (whatever that might be), but rather to convey useful information of some kind.

Modern computer rendering software takes its basic data from various kinds of standard structural databases which are compiled either from experimental X-ray scattering data, or are calculated from theory.

As was mentioned above, it is often desirable to show the "molecular surface"—the veil of negative charge that originates in the valence electrons of the atoms but which tends to be spread over the entire molecule to a distance that can significantly affect*van der Waals interactions* with neighbouring molecules.

Properties of Bonded Atom Pairs

When we talk about the properties of a particular chemical bond, we are really discussing the relationship between two adjacent atoms that are part of the molecule. *Diatomic* molecules are of course the easiest to study, and the information we derive from them helps us interpret various kinds of experiments we carry out on more complicated molecules.

It is important to bear in mind that the exact properties of a specific kind of bond will be determined in part by the nature of the other bonds in the molecule; thus the energy and length of the C–H bond will be somewhat dependent on what other atoms are connected to the carbon atom. Similarly, the C-H bond length can vary by as much a 4 per cent between different molecules. For this reason, the values listed in tables of bond energy and bond length are usually averages taken over a variety of compounds that contain a specific atom pair..

The energy of a system of two atoms depends on the distance between them. At large distances the energy is zero, meaning "no interaction". At distances of several atomic diameters attractive forces dominate, whereas at very close approaches the force is repulsive, causing the energy to rise. The attractive and repulsive effects are balanced at the minimum point in the curve. Plots that illustrate this relationship are known as *Morse curves,* and they are quite useful in defining certain properties of a chemical bond.

The internuclear distance at which the potential energy minimum occurs defines thebond length. This is more correctly known as the *equilibrium* bond length, because thermal motion causes the two atoms to vibrate about this distance. In general, the stronger the bond, the smaller will be the bond length.

Attractive forces operate between all atoms, but unless the potential energy minimum is at least of the order of *RT,* the two atoms will not be able to withstand the disruptive influence of thermal energy long enough to result in an identifiable molecule. Thus we can say that a chemical bond exists between the two atoms in H_2. The weak attraction between argon atoms does not allow Ar_2 to exist as a molecule, but it does give rise to the *van Der Waals force* that holds argon atoms together in its liquid and solid forms.

Potential energy and kinetic energy Quantum theory tells us that an electron in an atom possesses *kinetic energy* K as well as *potential energy* P, so the total energy E is always the sum of the two: $E = P + K$. The relation between them is surprisingly simple: $K = -0.5\ P$. This means that when a chemical bond forms (an exothermic process with $\Delta E < 0$), the decrease in potential energy is accompanied by an increase in the kinetic energy (embodied in the momentum of the bonding electrons), but the magnitude of the latter change is only half as much, so the change in potential energy always dominates. The *bond energy* ΔE has half the magnitude of the fall in potential energy.

BOND STRETCHING AND INFRARED ABSORPTION

When an atom is displaced from its equilibrium position in a molecule, it is subject to a restoring force which increases with the displacement. A spring follows the same law (Hooke's law); a chemical bond is therefore formally similar to a spring that has weights (atoms) attached to its two ends. A mechanical system of this kind possesses a natural vibrational frequency which depends on the masses of the weights and the stiffness of the spring. These vibrations are initiated by the thermal energy of the surroundings; chemically-bonded atoms are never at rest at temperatures above absolute zero.

Actual infrared spectra are complicated by the presence of more complex motions (stretches involving more than two atoms, wagging, etc.), and absorption to higher quantum states (overtones), so infrared spectra can become quite complex. This is not necessarily a disadvantage, however, because such spectra can serve as a "fingerprint" that is unique to a particular molecule and can be helpful in identifying it. Largely for this reason, infrared spectrometers are standard equipment in most chemistry laboratories.

The aspect of bond stretching and bending frequencies that impacts our lives most directly is the way that some of the gases of the atmosphere absorb infrared light and thus affect the heat balance of the Earth. Owing to their symmetrical shapes, the principal atmospheric components N_2 and O_2 do not absorb infrared light, but the minor components water vapour and carbon dioxide are strong absorbers, especially in the long-wavelength region of the infrared. Absorption of infrared light by a gas causes its temperature to rise, so any source of infrared light will tend to warm the atmosphere; this phenomenon is known as the greenhouse effect.

The incoming radiation from the Sun (which contains relatively little long-wave infrared light) passes freely through the atmosphere and is absorbed by the Earth's surface, warming it up and causing it to re-emit some of this energy as long-wavelength infrared. Most of the latter is absorbed by the H_2O and CO_2, the major greenhouse gasis in the unpolluted atmosphere, effectively trapping the radiation as heat. Thus the atmosphere is heated by the Earth, rather than by direct sunlight. Without the "greenhouse gases" in the atmosphere, the Earth's heat would be radiated away into space, and our planet would be too cold for life.

Radiation balance of the Earth In order to maintain a constant average temperature, the quantity of radiation (sunlight) absorbed by the surface must be exactly balanced by the quantity of long-wavelength infrared emitted by the surface and atmosphere and radiated back into space. Atmospheric gases that absorb this infrared light (depicted in red on the right part of this diagram) partially block this emission and become warmer, raising the Earth's temperature. This diagram is from the U. of Oregon.

Since the beginning of the Industrial Revolution in the 19th century, huge quantities of additional greenhouse gases have been accumulating in the atmosphere. Carbon dioxide from fossil fuel combustion has been the principal source, but intensive agriculture also contributes significant quantities of methane (CH_4) and nitrous oxide (N_2O) which are also efficient far-infrared absorbers. The measurable increase in these gases is believed by many to be responsible for the increase in the average temperature of the Earth that has been noted over the past 50 years- a trend that could initiate widespread flooding and other disasters if it continues.

IONIC AND COVALENT CHEMICAL BONDS

Chemical salts are found to conduct electricity when melted or dissolved in water, suggesting the presence of charge carriers, called ions (from the Greek for traveller). Since salts are neutral compounds, the charge carriers must be divided into two groups, one positively charged and the other negatively charged. The oppositely charged ions would be bound through electrostatic attractions, or ionic bonds. Because chemical salts are known to be composed of metallic elements bound to non-metallic elements, the most reactive of which are found at opposite ends of the periodic table, ionic attraction can be explained in terms of the attraction of opposites.

When metals or nonmetals bond to elements of their own kind, the reason for the bonding is not so obvious. Compounds containing different elements might have some attraction due to the differences between the elements, but homonuclear (one-element) polyatomic (more than one atom) elements such as H2 or sodium metal are bound by some common feature rather than by some difference between the atoms. Such bonds, called covalent bonds (Latin co + valere for joint strength), are not limited to homonuclear molecules but also occur between atoms of different elements that are close in behaviour. The periodic table summarizes the properties of the elements, including bonding trends as well, and can be used to decide the basic type of bonding between atoms in molecules in a binary.

Ionic Bonds

Ionic bonds are simply the attractive forces between oppositely-charged ions. Ionic compounds contain cations (which can be either monatomic or polyatomic) and anions (which can be either monatomic or polyatomic.) Remember that ions are formed by the loss or gain of electrons. For monatomic cations, the electrons which are lost first in the formation of the ion are always the outermost (valence) electrons. For example, you already know that the 4s subshell is at lower energy than the 3d subshell in a neutral atom. This is why the 4s subshell fills before the 3d. For ions, however, the order of orbital energies approaches that for a hydrogen ion, where all subshells in a given main shell have the same energy. For this reason, the 4s electrons are lost before the 3d electrons when a cation is formed from a transition metal. For example, the configuration of a neutral cobalt atom is [Ar]4s23d7, but the configuration of a cobalt(III) ion is [Ar]3d6. A simple view of the formation of an ionic compound is that the electrons lost by the cation are gained by the anion; in other words, a formal transfer of electrons can be envisioned. This is why the total positive charge from the cations must be balanced by the total negative charge from the anions - no free electrons are floating around. This gives rise to the formula Al_2O_3 for aluminum oxide rather than AlO_3, AlO, Al_3O_2, or other incorrect formulas.

DRAWING ELECTRON-DOT FORMULAS

- Sum up the total number of valence electrons for the elements in the molecule. If the molecule is charged, subtract one electron for each positive charge or add one electron for each negative charge.
- Write the structure for the molecule with a pair of electrons (or a dash) between each atom.

Groups of atoms will usually have the less electronegative atom surrounded by atoms having greater electronegativity. Never place a hydrogen atom in the centre since it can only form one bond.

Think about the valence of each atom and make sure that these valences are not exceededexcept in the case of ions. Ions always have more or less bonds than the normal uncharged atom.

The common valence or number of bonds formed for some common atoms are:

H = 1

O and S = 2

N = 3

C = 4

F, Cl, Br and I = 1

- Place electrons around the outer atoms to fill their outer shells. Most atoms require eight electrons ("octet rule") so they will resemble an inert gas. Elements in groups IA to IIIA do not follow the octet rule and have less than eight electrons in the final formula.
- Subtract the number of electrons used so far from the total calculated in step 1 and place these remaining electrons on the central atom or atoms.
- If the central atom ends up with less than 8 electrons, then it probably forms a multiple bond with an adjacent outer atom.
- Finally you should calculate the formal charge on the atoms.

Exceptions to the Octet Rule

It is very important that you understand the basics of what we have just covered before worrying about exceptions. In almost all cases, the elements in the first and second periods are used for electron-dot formulas. There are important exceptions but these are not as common. Elements in groups IA, IIA and IIIA do not follow the "octet rule" that we used for the above problems. When we write the electron-dot formula for BF_3, the boron will not have eight electrons and that is just fine...it helps us understand its chemical character.

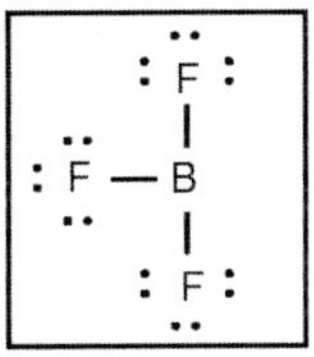

Some elements having low-energy d-orbitals also form exceptions to the "octet rule" in that more than eight electrons are accommodated around the central atom. The central atom in most of these compounds will be bonded to highly electronegative elements such as fluorine, oxygen and chlorine. A surprising element in this group is the inert gas, xenon. If xenon is exposed to fluorine gas in the presence of light for several weeks it can form XeF_2, a colourless crystalline solid.

Let's try to draw the electron-dot formula for SF_4 (sulfur tetrafluoride). Note that sulfur is in the 3rd period and thus does have d-orbitals available. Transition metals offer a unique problem in that they have several common oxidation states so we tend to not write electron-dot formulas for their compounds.

Resonance

Resonance theory is one of the most important theories that helps explain many interesting aspects of chemistry ranging from differences in reactivity of related compounds to physical properties such a the absorption of light by molecules. The electron-dot formula for many of the compounds and ions presented us a choice when we placed 4 electrons between 2 of the atoms in the formation of double bonds. The carbonate and nitrate anions are examples of this problem. You may have wondered why your structure differed from the structure drawn in this tutorial in where the double bond was located. It turns out that it doesn't matter!

The same concept holds true for nitric acid and, in this case, the charge is evenly distributed among the 3 oxygen atoms in the nitrate anion. This makes the negative charge less available for the reverse reaction and helps explain why nitric acid is a fairly strong acid. The bond order for the nitrogen-oxygen bonds in the nitrate anion is 1.33.

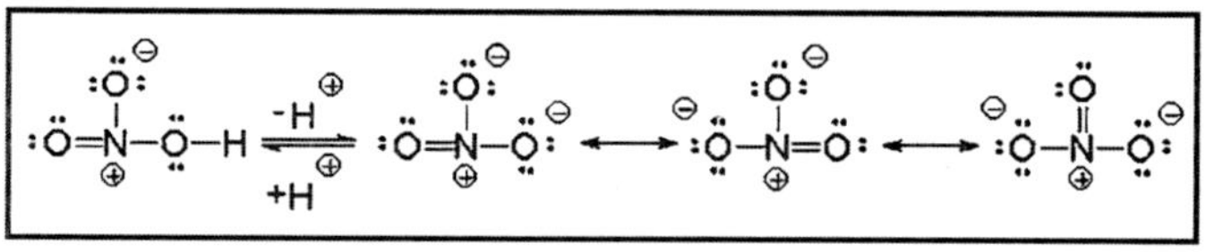

BASIC BONDING CONCEPTS

The most basic concept in all of chemistry is the chemical bond. The chemical bond is essentially the sharing of electrons between two atoms, a sharing which holds or bonds the atoms together. In order to understand the chemical bond, atoms and electrons must first be understood.

Electron Orbitals

Atoms have three components: protons, neutrons, and electrons. Protons have a positive charge of +1, neutrons have 0 charge, and electrons have a negative charge of -1. The protons and neutrons occupy the centre of the atom as a piece of solid matter called the nucleus. The electrons exist in "orbitals" surrounding the nucleus. In the olden times, it was believed that the electrons orbited the nucleus in the same way that planets orbit the sun. This is not exactly true, but it explains the origin of the term orbital to describe the location of the electron. In reality, it is impossible to tell the precise trajectory of an electron in the way that the path of a planet can be followed. The best that can be achieved is to describe the probability of locating the electron in a region of space. The distribution of this probability over space is called an orbital. Let us treat the simplest case, in which the nucleus is surrounded by just one electron (for example, the hydrogen atom). In this case, the probability of finding an electron in its lowest energy, or most stable, state is distributed in a spherically symmetric way around the nucleus. The probability of finding the electron is highest at the nucleus and decreases as the distance from the nucleus increases.

The density of dots in a region represents to the probability of locating the electron there. This lowest energy, spherically symmetric orbital is called the 1s orbital. As stated, in this orbital, the highest probability of finding the electron at a point in space occurs at the nucleus. Such a location may seem counterintuitive, since the electron is known to generally be located some distance from the nucleus, not directly at it. How can this apparent contradiction be explained?

Although the probability of finding the electron at a point is greatest at the nucleus, the volume of space occupied by the nucleus is extremely small compared to the rest of the atom. The probability of finding the electron at a point decreases farther from the nucleus, but the volume in which it can be found becomes much, much larger. So the electron is more likely to be found some distance from the nucleus. It is the product of the volume in which the electron can be found and the probability of finding it that gives the probability of finding the electron as a function of distance from the nucleus. The radial distribution function, which is a product of the probability and the volume element, gives the probability of finding the electron in a spherical shell some distance r from the nucleus. It is clear that the electron is most likely to be found in a shell some distance from the nucleus rather than directly at the nucleus. In fact, since the volume element is almost zero directly at the nucleus, the electron is very unlikely to be found there.

The 1s orbital is the lowest energy orbital that an electron can occupy, but several higher energy orbitals are significant in organic chemistry. The next lowest energy orbital that an electron can occupy is the 2s orbital, which looks much like the 1s orbital except that the electron is more likely to be found farther from the nucleus. The third lowest energy orbital is called the 2p orbital. The major and highly important difference between a p orbital and an s orbital is that the p orbital is not spherically symmetric. Instead, the p orbital is oriented along a specific axis in space. There are three p orbitals, oriented along the x, y, and z axes.

So far, our discussion of orbitals has been mathematically exact. The images of s and p orbitals depict the actual distribution over space of the probability of finding an electron in atoms with one electron. However, after a second electron is added, the problem cannot be solved exactly due to the electron-electron repulsion. Instead, approximations have to be made. Fortunately, using the exact orbitals of the one-electron atom as a basis for the orbitals in many-electron atoms is an excellent approximation. For example, in carbon, the energy of the electron is correct to within 1% using this approximation. Thus, orbitals in many-electron atoms look essentially like the one-electron atom orbitals that have already been described.

COVALENT BOND ENERGIES

Bonds are stable because the ions in an ionic bond are at lower energy when they are close to one another in the crystal, and because the shared electrons in a covalent bond are attracted by the two nuclei of the bonded atoms. Thus, it takes energy to break ionic and covalent bonds. Covalent bond energies are average energies, each of which depends on the chemical environment of the bonded atom. For example, the successive bond dissociation energies for methane are shown on page 366 of the text. Although the four values (one for each C-H bond) are slightly different, their average value of 413 kJ/mol can be assumed to represent the energy of a covalent carbon-hydrogen bond. A number of bond energies are shown in Table. Notice that the table is separated into single bond and multiple bond categories. Table lists bond lengths for selected bonds. Notice that for a given pair of atoms, bond length decreases in the order,

single > double > triple.

Since the formation of covalent bonds releases energy and the breaking of bonds requires energy, we can calculate the approximate the enthalpy change for a chemical reaction by using tabulated bond energies. The procedure is simple: first, break apart all the bonds in the reactants and determine how much energy is required (this is endothermic, and thus a positive value.) Then determine how much energy is released when the bonds in the products form (this is exothermic, and thus a negative value.) Add the two quantities to get the enthalpy change for the reaction.

Important note: It is essential that the type of bonds in the reactants and products are known. For example, simply looking at the formula N2 might give one the impression that a single-bond N-N energy of 160 kJ per mole should be used. However, this will result in an incorrect answer, since there is a triple bond in the nitrogen molecule with a covalent bond energy of 941 kJ/mole. The types of bonds in a molecule can be determined by drawing the Lewis electron-dot structure for the molecule.

Resonance

Although Lewis structures are very useful for a large number of molecules, there are some problems with using electron-dot structures to describe bonding:

One of the goals in drawing a Lewis structure is to pair up electrons. This is impossible for a molecule with an odd number of electrons, like NO or ClO_2. Which atom "has" the odd electron in these molecules?

Another goal for obtaining a valid Lewis structure is to obtain a duet of electrons around hydrogen or a completed octet around other atoms. The fact, however, is that there are exceptions to the octet rule: Be in BeF2 and B in BF3 have only four and six electrons in the Lewis structure, respectively, instead of eight. Atoms in the third period and below are able to "expand" their octet and can thus have more than eight electrons in the Lewis structure. Some examples are SF4 (10 e- around sulfur) and XeF4 (12 e- around xenon). There are many more examples.

Sometimes a single Lewis structure for a molecule doesn't accurately explain the properties of the molecule. For example, the Lewis structure for sulfur dioxide, SO_2, has a single sulfur-oxygen bond, a double sulfur-oxygen bond, and a lone pair of electrons on the central sulfur atom. All electrons are paired, 18 valence electrons are used in the Lewis structure, and each atom has an octet.

Thus, the Lewis structure is "correct", at least in a formal sense. The problem is that the electron-dot diagram implies that there should be two bond strengths in sulfur dioxide - an S-O single bond and a stronger S-O double bond. Furthermore, there should also be two different bond lengths - a longer S-O single bond and a shorter S-O double bond. Here's the problem: experimentally, it has been found that both sulfur-oxygen bonds have the same bond length and the same bond strength. The concept of resonance is used to overcome this sort of problem with Lewis structures.

The problem mentioned above is not a problem with the molecule - it is a problem with the limited ability of a simple notation like a Lewis structure to describe a complicated concept like the bonding in a molecule. It is rather like trying to describe a real, three-dimensional object by drawing it on a two-dimensional piece of paper; the problem isn't with the object - it is with our limited resources for making the drawing.

The best way to describe the bonding in molecules like SO_2 is to draw two or more Lewis structures and say that the actual molecule is a hybrid, or combination, of the separate structure; this is indicated by drawing double-headed arrows between the structures.

Each of the separate structures is called a resonance structure, and the actual molecule is described as a resonance hybrid of the resonance structures.

It is important to remember two rules for drawing resonance structures: The positions of all atoms in all resonance structures of a given molecule must be the same. In other words, atoms are not moved when a different resonance structure is drawn from an existing one; rather, electron pairs are moved. Do not think that the actual structure results from an oscillation among the resonance structures. For example, each of the two resonance structures for sulfur dioxide has a single bond and a double bond: the difference is in the location of each.

Don't think that at a particular instant there is a single bond between the sulfur and one of the oxygens and that in the next instant the bond becomes a double bond, with continual "flipping" back and forth. This is not the correct interpretation of resonance. Both bonds in SO_2 are identical, and the bonds are neither single nor double bonds. Each bond has properties intermediate between a single and double bond. The resonance hybrid can be compared to a mule, which is a hybrid between a horse and a donkey. The mule has its own unique properties, and it exists in its own right — it isn't a horse half the time and a donkey the other half of the time.

Formal Charge

Unlike oxidation number, which can be determined from a formula like Na_2SO_4, the formal charge of an atom in a covalent molecule or polyatomic ion can only be determined from the Lewis structure. Formal charge is calculated by splitting the bonding electrons equally between the bonded atoms, and assigning all nonbonding electrons to the atom which "owns" them, then comparing the resulting number of electrons with the number of valence electrons the atom would have if it were neutral. Formal charges are shown as a positive number or negative number (*e.g.* +2 or –1) inside a circle and placed next to the atomic symbol in the Lewis structure. Recall that oxidation numbers are written without the circle.

For example, the Lewis structure for carbon dioxide shows two carbon-oxygen double bonds and two nonbonding pairs of electrons on each of the oxygen atoms. The carbon atom has four bonding pairs of electrons (two double bonds, or eight electrons) around it. The central carbon atom is assigned four of these eight electrons and each oxygen is assigned two of them. The formal charge of carbon is therefore number of valence electrons minus number of electrons after assignment, or 4 minus 4 = a formal charge of zero. Each oxygen atom "owns" two nonbonding pairs of electrons (four electrons) and was assigned two of the bonding electrons, giving a total of six electrons. The formal charge is therefore 6-6= a formal charge of zero. Now draw one of the resonance structures for sulfur dioxide, with one S-O single bond, one S-O double bond, and a nonbonding pair on the central sulfur atom (remember to complete the octets around the oxygen atoms.) Assigning the electrons leads to a formal charge of +1 for the sulfur, -1 for the singly-bonded oxygen and zero for the doubly-bonded oxygen.

Let's look at one more example. Draw the Lewis structure for the sulfate ion, $(SO_4)^{2-}$, which has four single sulfur-oxygen bonds. Verify the formal charges of +2 for the central sulfur atom and -1 for each of the oxygen atoms. Notice that in the CO_2, SO_2, and $(SO_4)^{2-}$ examples above, the algebraic sum of the formal charges is equal to the charge on the molecule or ion. This must always be the case. Remember that formal charges cannot be determined from a simple formula; the Lewis structure is needed.

Formal Charges and Resonance

Formal charges can be used to decide if a particular resonance structure is a "good" one - in other words, if the resonance structure contributes significantly to the actual structure.

Here are the rules:

- "Like" formal charges (plus/plus or minus/minus) should not reside on adjacent atoms.
- "Unlike" formal charges (plus/minus) should not be unnecessarily separated.

Formal charges closer to zero usually represent more important resonance structures than those which have large formal charges. For example, draw one of the Lewis structures for nitrous oxide (laughing gas, N_2O - the skeletal structure is N-N-O). You should now be able to "push" electron pairs around to come up with a total of three resonance structures which obey the octet rule. Each of these resonance structures. Make sure you correctly determined these formal charges for all three of the resonance structures (Formal charges are listed in the following order: terminal nitrogen, central nitrogen, oxygen.)

- N-N double bond, N-O double bond: formal charges are –1, +1, 0.
- N-N triple bond, N-O single bond: formal charges are 0, +1, and -1.
- N-N single bond, N-O triple bond: formal charges are –2, +1, and +1.

The resonance structures 1 and 2 are "good" structures and contribute significantly to the actual structure. Resonance structure 3 results in the placement of two positive formal charges on adjacent atoms, and thus does not contribute significantly to the actual structure.

BONDING IN CARBON

That a chemical bond is essentially the sharing of electrons between two atoms. Since electrons are negatively charged and exert an attractive force on nuclei, they serve to hold the atoms together if they are located between two nuclei. How does this notion of a chemical bond relate to orbitals describing electron density? When two atoms approach each other, their atomic orbitals overlap. The overlapped atomic orbitals can add together to form molecular orbitals. This procedure is called the linear combination of atomic orbitals (LCAO). The area of greatest overlap between the original atomic orbitals represents the chemical bond that is formed between them. Since the sharing of electrons is the basis of the chemical bond, the molecular orbitals formed represent chemical bonds.

Let us take a simple example, hydrogen. We can imagine two 1s orbitals gradually coming closer and closer together until there is a good deal of overlap between them. At this point, the area in space of greatest electron density will be between the two nuclei, which themselves were at the centre of the original atomic orbitals. This electron density, now part of a new molecular orbital, represents the chemical bond. When, as below, the area of greatest overlap occurs directly between the two nuclei on internuclear axis or axis containing the nuclei of both atoms, the bond is called a sigma bond or σ bond.

There are two final points to consider when atomic orbitals combine. First, more than one atomic orbital from a single atom can be used to form new molecular orbitals. For example, a 2s orbital and a 2p orbital from one atom might add together and overlap with one or more orbitals from a second atom to form new molecular orbitals. Second, parts of orbitals can possess a sign (+ or -). The s orbital has the same sign throughout, while in the p orbitals, one lobe is + and the other lobe is -. Signs do not matter with respect to electron density, but they must be taken into account when orbitals are added or subtracted. If two orbitals of the same sign are added, electron density will increase, while if two orbitals of opposite signs are added, the shared electron density will cancel out.

sp^3 Bonds

Carbon has six electrons. Only two electrons can occupuy an s orbital at a time. The first two electrons in carbon occupy the 1s orbital. The next two occupy the higher-energy, but similarly shaped 2s orbital. The final two electrons occupy the 2p orbitals. In carbon, the electrons in the 1s orbital are too low in energy to form bonds. Thus, electrons used to form bonds must come from the 2s and 2p orbitals. Carbon very often makes four bonds, and when it does so, these bonds are arranged so that they are as far away from each other as possible. This arrangement is referred to as tetrahedral.

It is clear that the individual 2s and 2p orbitals cannot form bonds in this arrangement due to their geometry. The 2s orbital is completely symmetric, while the 2p orbitals are aligned along specific axes. None of these orbitals is well-equipped to form bonds in the tetrahedral geometry alone. However, we have stated that a chemical bond doesn't have to be formed from individual atomic orbitals, but can be formed from a combination of several atomic orbitals from the same atom. For each bond that is made in the tetrahedral geometry, a part of the 2s and a part of each of the 2p orbitals will contribute, resulting in a tetrahedral arrangement. To achieve this geometry, both the 2s and all three of the 2p orbitals ($2p_x$, $2p_y$, and $2p_z$) must contribute. The new bonds that are formed are called sp^3 bonds, since one s and 3 p orbitals were used to form them.

Note that when sp^3 bonds are formed, they all point in separate directions as far away from each other as possible, leading to 109.5° angles between the bonds. This is the tetrahedral geometry as shown. When carbon makes four sp^3 bonds to hydrogen, the resulting molecule is tetrahedral in shape. Carbon can also make a bond to another carbon atom. Two carbon atoms, both of which have sp^3 bonds oriented tetrahedrally, can bond to each other if each uses one of its sp^3 bonds.

sp^2 Bonds

Carbon sometimes makes just three bonds. In this case, not all of the 2p orbitals combine with the 2s orbital to form bonds. Instead, a combination of the 2s orbital and two of the 2p orbitals make three sp^2 bonds, while the other p orbital does not participate in this combination and can make a fourth bond on its own.

Like the sp^3 bonds, the sp^2 bonds are oriented such that they are as far away from each other as possible. This geometry is called the trigonal planar geometry. It is called trigonal planar since each of the bonds points to one of the vertices of a triangle, but all three bonds are located in the same plane. What about the other 2p orbital, the one which did not add to make sp^2 bonds? This orbital still exists, and it is perpendicular to the plane in which the sp^2 bonds form. It too is able to form bonds, and it does so independently of the sp^2 bonds.

When two carbon atoms with sp^2 orbitals form a bond to each other using their sp^2 orbtials, a σ bond is formed between then. Morever, the extra p orbtials that exist above and below each carbon atom also overlap with each other. This orverlap between p orbitals leads to the formation of a second bond in addition to the σ bond formed between the sp^2 orbitals. This second bond, which does not occur directly between the nuclei on the internuclear axis, but rather above and below the internuclear axis, is called a pi bond or π bond. When a σ bond and a π bond form together between two atoms, a double bond is said to have formed.

IONIC BONDING

Radius Ratio Rules

For the stability of an ionic compound, each cation should be surrounded by maximum number of anions and vice versa. The number of oppositely charged ions surrounding each ion is called its co-ordination number.

The larger is the size of the cation the greater is its coordination number. In other wards, greater is the

$$\text{radius ration} = \left(\frac{\text{Radius of the cation}}{\text{Radius of the anion}}\right)$$

the greater is its coordination number.

The relationship between the radius and the coordination number and the structural arrangement are called radius ratio rules.

The radius ratio rule is given in the table below.

$$\text{Radius ratio} = \frac{\text{Radius of cation } (r^+)}{\text{Radius of anion } (r^-)} = \frac{r^+}{r^-}$$

Radius ratio	*Co-ordination no.*	*Structure*
0.155-0.225	3	Plane triangular
0.225 - 0.414	4	tetrahedral
0.414 - 0.732	6	Octahedral
0.732 -1	8	Body - centred cubic

The above relation may be understood with the following explanation.

For a most stable arrangement the anions must touch each other as well as the cation simultaneously.

Similarly in case of an octahedral arrangement, if the size of the cation decreases i.e., the radius ratio decreases the arrangement changes to tetrahedral to become stable. But if the size of the cation increases i.e, the radius ration increases the arrangement changes to body centred cubic structure to acquire stability.

LATTICE ENERGY

The lattice energy of an ionic solid is a measure of the strength of bonds in that ionic compound. It is usually defined as the enthalpy of formation of the ionic compound from gaseous ions and as such is invariably exothermic. The concept of lattice energy was initially developed for rocksalt-structured and sphalerite-structured compounds like NaCl and ZnS, where the ions occupy high-symmetry crystal lattice sites. In the case of NaCl, the lattice energy is the energy released by the reaction

$Na^+ (g) + Cl^- (g) \rightarrow NaCl (s)$

which would amount to -787 kJ/mol.

Some older textbooks define lattice energy as the energy required to convert the ionic compound into gaseous ions which is an endothermic process, and following this definition the lattice energy of NaCl would be +787 kJ/mol. The precise value of the lattice energy may not be determined experimentally, because of the impossibility of preparing an adequate amount of gaseous anions and cations and measuring the energy released during their condensation to form the solid. However, the value of the lattice energy may either be derived theoretically from electrostatics or from a thermodynamic cycling reaction, the so-called Born-Haber cycle.

THEORETICAL TREATMENTS

Born-Landé Equation

In 1918 Born and Landé proposed that the lattice energy could be derived from the electric potential of the ionic lattice and a repulsive potential energy term.

$$E = -\frac{N_A M z^+ z^- e^2}{4\pi\varepsilon_0 r_0}\left(1 - \frac{1}{n}\right),$$

where

N_A is the Avogadro constant;

M is the Madelung constant, relating to the geometry of the crystal;

z^+ is the charge number of cation;

z^- is the charge number of anion;

e is the elementary charge, equal to 1.6022×10–19 C;

ε_0 is the permittivity of free space, equal to 8.854×10–12 $C^2\ J^{-1}\ m^{-1}$;

r_0 is the distance to closest ion; and

n is the Born exponent, a number between 5 and 12, determined experimentally by measuring the compressibility of the solid, or derived theoretically. The Born-Landé equation gives a reasonable fit to the lattice energy.

Compound	*Calculated Lattice Energy*	*Experimental Lattice Energy*
NaCl	–756 kJ/mol	–787 kJ/mol
LiF	–007 kJ/mol	–1046 kJ/mol
$CaCl_2$	–2170 kJ/mol	–2255 kJ/mol

From the Born-Landé equation it can be seen that the lattice energy of a compound is dependent on a number of factors:

- as the charges on the ions increase the lattice energy increases (becomes more negative),
- when ions are closer together the lattice energy increases (becomes more negative)

Barium oxide (BaO), for instance, which has the NaCl structure and therefore the same Madelung constant, has a bond radius of 275 picometers and a lattice energy of -3054 kJ/mol, while sodium chloride (NaCl) has a bond radius of 283 picometers and a lattice energy of -786 kJ/mol.

Kapustinskii Equation

The Kapustinskii equation can be used as a simpler way of deriving lattice energies where high precision is not required.

Effect of Polarisation

For ionic compounds with ions occupying lattice sites with crystallographic point groups C_1, C_{1h}, C_n or C_{nv} ($n = 2, 3, 4$ or 6) the concept of the lattice energy and the Born-Haber cycle has to be extended. In these cases the polarization energy E_{pol} associated with ions on polar lattice sites has to be included in the Born-Haber cycle and the solid formation reaction has to start from the already polarized species. As an example, one may consider the case of iron-pyrite FeS_2, where sulfur ions occupy lattice site of point symmetry group C_3. The lattice energy defining reaction then reads

Fe^{2+} (g) + 2 pol S^- (g) → FeS_2 (s)

where pol S^- stands for the polarized, gaseous sulfur ion.

It has been shown that the neglection of the effect led to 15 % difference between theoretical and experimental thermodynamic cycle energy of FeS_2 that reduced to only 2%, when the sulfur polarization effects were included.

BORN-HABER CYCLE

The Born–Haber cycle is an approach to analysing reaction energies.

It was named after and developed by the two German scientists Max Born and Fritz Haber.

The Born–Haber cycle involves the formation of an ionic compound from the reaction of a metal (often a Group I or Group II element) with a non-metal.

Born–Haber cycles are used primarily as a means of calculating lattice energies (or more precisely enthalpies) which cannot otherwise be measured directly.

The lattice enthalpy is the enthalpy change involved in formation of the ionic compound from gaseous ions. Some chemists define it as the energy to break the ionic compound into gaseous ions. The former definition is invariably exothermic and the latter is endothermic.

A Born–Haber cycle calculates the lattice enthalpy by comparing the standard enthalpy change of formation of the ionic compound (from the elements) to the enthalpy required to make gaseous ions from the elements. This is an application of Hess's Law.

It is this latter calculation that is complex. To make gaseous ions from elements it is necessary to atomise the elements (turn each into gaseous atoms) and then to ionise the atoms. If the element is normally a molecule then we have to consider its bond dissociation enthalpy. The energy required to remove one or more electrons to make a cation is a sum of successive ionization energies; for example the energy needed to form Mg^{2+} is the first plus the second ionization energies of Mg. The energy released when one electron is added to an atom to make it an anion is called the electron affinity.

4

Elementary Processes and Reaction Mechanism

A balanced chemical equation is intended to represent the net transformation of reactants to products that occurs when reaction takes place. It does not, and is not intended to, tell us anything about the details of the reaction pathway. In fact, most reactions occur via a series of relatively simple steps, each involving reaction of only a few (one, two, and very occasionally, three) molecules. Each of these simple steps is called an elementary process. The equation for an elementary process represents an actual collision event at the molecular level. The sequence of elementary processes by which a reaction is proposed to occur is called the mechanism for the reaction. For example, the reaction is thought to occur by the 2-step mechanism.

Overall:

$$2NO_2\,(g) + F_2\,(g) \rightarrow 2NO_2F\,(g)$$

Mechanism:

$$NO_2 + F_2 \rightarrow NO_2F + F$$
$$NO_2 + F \rightarrow NO_2F$$

with the first step proposed to be slow, the second one fast. Each of the two steps in the mechanism is understood to be an elementary process. We can make several important generalizations about the mechanism. First, each elementary process involves reaction (collision) of only two species, even though the overall reaction involves 3 moles of reactants. Second, F atoms, produced in the first step and consumed in the second, do not appear in the overall equation for the reaction. Such a species—first produced, then consumed—is called a reaction intermediate. Third, the elementary processes add to give the overall equation for the reaction. This must always be true.

It is important to realise that a mechanism for a reaction is a hypothesis about what goes on at the molecular level, made by the chemist based on her or his experimental observations. The overall equation for a reaction is NOT a hypothesis; it is an experimental fact, observable in the laboratory. The mechanism in all of its detail is seldom observable; it must be deduced from experiment, must be consistent with all known experimental facts about the reaction, and is subject to change if in conflict with a new experimental fact.

Types of Elementary Processes

There are three types of elementary process, named for their molecularity—the number of reactant molecules that they involve. A unimolecular process involves only one reactant molecule. Two examples of unimolecular processes are given.

$$O_3 \rightarrow O_2 + O$$
$$I_2 \rightarrow 2I$$

A bimolecular process involves two reactant molecules, which may be the same or different. Examples are given. Bimolecular elementary processes are by far the most common type.

$$NO_2 + F \rightarrow NO_2F$$
$$H_2O_2 + I^- \rightarrow H_2O + IO^-$$

A termolecular process requires the simultaneous collision of three reactant molecules. An example.

$$2NO + O_2 \rightarrow N_2O_4$$

Such three-body collisions are extremely improbable. Consequently, termolecular processes are rare. Elementary processes with molecularity greater than three do not occur, due to the overwhelming improbability of collisions involving four or more molecules. In contrast to the situation for the overall (net) reaction, the rate law for an elementary process may be written directly from its stoichiometry, because an elementary process is an actual, not a net, process. Thus for the elementary process,

$$A + B \rightarrow C + D$$

the rate law is $-D[A]/dt = k[A][B]$. Unimolecular processes have first-order rate laws. The rate law is

$$-D[O_3]/dt = k[O_3]$$

Bimolecular processes have rate laws that are second-order overall; and termolecular processes have overall third order rate laws.

$$-D[NO_2]/dt = k[NO_2][F]$$
$$-D[O_2]/dt = k[NO]^2[O_2]$$

In summary, an elementary process of molecularity n has a rate law of order *n*. The converse is NOT true: a first-order reaction is not necessarily either an elementary process or unimolecular.

The difference between order and molecularity is subtle but important. The concept of molecularity is appropriate only in the context of an elementary process, and indicates the number of reactant molecules involved in (undergoing collision) in the elementary step. The order of a reaction is the number to which the concentration of a reactant must be raised in the experimental rate law for the overall (net) reaction. This rate law, and the orders in it, do not necessarily correspond to a particular elementary step in the mechanism of the net reaction. We will say more about this shortly.

Rate Determining Step

A particular overall reaction consists of a sequence of elementary steps, which usually occur with different rates. Very often one of the elementary processes occurs much more slowly than do the others and limits the rate at which the overall reaction occurs. This slow step is then referred to as the rate-determining step (rds). The overall rate is equal to the rate of the rds. The fact that the overall reaction rate is limited by the rate of a particular elementary step is the reason why rate laws frequently do not correspond with the stoichiometry of the overall reaction.

A familiar example of an overall process occuring at a rate limited by a slow step. Theater filled to capacity, connected by a set of wide double doors to a smaller lobby, which is linked to the outside via a revolving door. When the movie is over, people begin to move from the theater to the lobby, and from there to the outside.

We suppose that the double doors between theater and lobby are large enough so that 10 people per second can pass through; but that only 1 person per second can exit the revolving door. We then represent the overall process of leaving the theater in terms of 2 steps:

$$\text{theater} \rightarrow \text{lobby [10/s]}$$

$$\text{lobby} \rightarrow \text{outside [1/s]}$$

$$\text{Overall, theater} \rightarrow \text{outside [Rate = ?]}$$

Initially, people pass from theater to lobby at the rate of 10 per second, and begin to exit the revolving door one by one. However, the lobby quickly fills up, because it cannot be emptied nearly as rapidly as it was filled. The process "piles up" behind the slow step, which is passage through the revolving door. The maximum rate at which people can leave the theater is 1 per second, which is the rate of the slowest step in the process.

Rate Law and Reaction Mechanism

Chemists are very interested in the details of how a reaction occurs; *i.e.*, in the reaction mechanism. The approach to the elucidation of the mechanism of a reaction of known stoichiometry involves at least the following steps:

- Do experiments to determine the experimental rate law for the overall reaction;

- Propose a mechanism for the overall reaction, based on the observed rate law;
- Derive the theoretical rate law implied by the proposed mechanism;
- Compare the rate laws in 1 and 3. If they have the same form, it is safe to conclude that the mechanism is consistent with (not in conflict with) the experimental rate law. If they do not have the same form, go back to 2.

Step 2 makes use of the fact that the observed rate law indicates what atoms are involved as reactants in the rate-determining step of the mechanism. This will be made clear in specific examples to follow. It is the most difficult step of the above procedure, requiring extensive experience and well-developed "chemical intuition." At this level of study, we will not attempt to deduce mechanism from the rate law. Rather, our focus will be this: given the form of the experimental rate law, and one or more proposed mechanisms, decide whether each mechanism is or is not consistent with experiment.

Example:. For the reaction $2NO_2(g) + F_2(g) \rightarrow 2NO_2F(g)$, the experimental rate law has the form, Rate = $k_{obs}[NO_2][F_2]$. Two mechanisms have been proposed. These are given below:

Mechanism 1:

$$NO_2 + F_2 \Leftrightarrow NO_2F + F \; [k_1, k_{-1} \text{ fast}]$$

$$NO_2 + F \rightarrow NO_2F \; [k_2 \text{ slow (rds)}]$$

Mechanism 2:

$$NO_2 + F_2 \rightarrow NO_2F + F \; [k_1 \text{ slow (rds)}]$$

$$NO_2 + F \rightarrow NO_2F \; [k_2 \text{ fast}]$$

Can either of these mechanisms be ruled out as being in conflict with experiment?

Solution: We examine mechanism 1 first. To obtain the theoretical rate law (implied rate law) for this mechanism, we proceed as follows:

- Write the rate law for the rds. The rate of this step is the same as the rate of the overall reaction.

The rds for mechanism 1 is step 2.

$$\text{Rate} = k_2[NO_2][F]$$

Although this rate law is valid, it is not expressed solely in terms of concentrations of the reactants in the overall reaction. Since these concentrations are the ones we actually observe, we must somehow rewrite the rate law so as to eliminate the concentration of the intermediate, F, which is not generally observable, and replace it with concentrations of reactants in the net reaction.

We can accomplish this making use of the preceding rapid steps, which we assume to occur rapidly enough to come to a pseudo equilibrium. So,

- Use the rapid pre-equilibrium to solve for the concentration of the intermediate in terms of known concentrations. Since the first step of the mechanism is in equilibrium, the forward and reverse rates of this elementary process are the same, and we can write,

$k_1[NO_2][F_2] = k_{-1}]NO_2F][F]$

Solving for [F] gives

$[F] = k_1/k_{-1}\ \{[NO_2][F_2]/[NO_2F]\}$

This expression for the concentration of the intermediate can be substituted into the rate law for the rds:

$Rate = k_1k_2/k_{-1}\ \{[NO_2]^2[F_2]/[NO_2F]\}$

This is the rate law implied by the mechanism. It is the rate law that the reaction should follow if the proposed mechanism is correct. Clearly, though, it is not in agreement with the experimental rate law, which is first order in NO_2, and shows no dependence on NO_2F. The conclusion is that Mechanism 1 can NOT be correct; it is in conflict with experiment. We next consider Mechanism 2, using the same general approach.

- rite the rate law for the rds, which in this case is the first step in the mechanism.

$$Rate = k_1[NO_2][F_2]$$

Further work is unnecessary here, because the rds rate law is already expressed in terms of concentrations of participants in the overall reaction. The implied rate law for Mechanism 2 has the same form as the experimental rate law (first order in NO_2, first order in F_2, second order overall), and is therefore not in conflict with experiment.

Can we conclude, then, that Mechanism 2 is the correct mechanism for reaction? The answer is an emphatic NO. If a mechanism does not agree with experiment, it must be wrong; but just because a mechanism agrees with experiment does not mean that it is correct. A mechanism is a theory—it can never be proved. To underscore this point, consider the following possible mechanism:

Mechanism 3:

$$NO_2 + F_2 \rightarrow NO_2F_2\ [k_1\ slow]$$
$$NO_2F_2 + NO_2 \rightarrow 2NO_2F\ [k_2\ fast]$$

This also is consistent with experiment, giving the implied rate law,

$$Rate = k[NO_2][F_2]$$

based on the first step, which is the slow one. You might like to assess the validity of yet a fourth possible mechanism, given below:

Mechanism 4:

$$1/2\ F_2 \rightarrow F\ [fast]$$

$$F + NO_2 \rightarrow NO_2F\ [slow]$$

We can now appreciate the significance of a statement made earlier: the experimental rate law tells us what atoms are contained in the reactants of the rate-determining step. The experimental rate law in Example., Rate = k $[NO_2][F_2]$, reveals that the reactants in the slow step must contain a total of 1 N atom, 2 O atoms, and 2 F atoms.

Based on this, we can immediately rule out mechanism 1, for which the rds involves only one F atom; and mechanism 4, for which the same is true. Both mechanisms 2 and 3 have rds's containing the correct numbers of atoms, so neither is in conflict with experiment.

In general, chemists try to propose mechanisms that agree with the experimental rate law and with any other experimental evidence; and that are chemically reasonable. It is the assessment of chemical reasonability that requires experience. We consider one additional example of mechanism assessment.

Example: Several mechanisms might be considered for the overall reaction

$$2NO + O_2 \rightarrow 2NO_2$$

with rate law, Rate = $-d[O_2]/dt = k_{obs}[NO]^2[O_2]$. The symbol, k_{obs}, where "obs" stands for "observed", is commonly used for an experimentally measured rate constant.

Mechanism 1:

$$NO + NO \Leftrightarrow N_2O_2 \; [K_1 \text{ fast}]$$

$$N_2O_2 + O_2 \rightarrow 2NO_2 \; [k_2 \text{ slow}]$$

Mechanism 2:

$$NO + O_2 \Leftrightarrow OONO \; [K_1' \text{ fast}]$$

$$NO + OONO \rightarrow 2NO_2 \; [k_2 \text{ slow}]$$

Mechanism 3:

$$NO + O_2 \rightarrow OONO \; [k_1 \text{ slow}]$$

$$NO + OONO \rightarrow 2NO_2 \; [k_2 \text{ fast}]$$

Can any of these mechanisms be ruled out?

Solution: Mechanism 3 can be eliminated right away, because the reactants in the rds do not contain the correct numbers of atoms.

The experimental rate law demands that there be two N atoms and four O atoms in the rds reactants; however, only one N and three O atoms are involved in the rds of mechanism 3. Mechanisms 1 and 2 show the correct atoms in the rds. You should confirm that the implied rate laws are,

$$\text{Mechanism 1: Rate} = k_2K_1[NO]^2[O_2]$$

$$\text{Mechanism 2: Rate} = k_2K_1'[NO]^2[O_2]$$

both of which are consistent with experiment. Attempts to draw Lewis structures for the intermediates, N_2O_2 and OONO. The intermediate OONO is a free radical (it has an odd electron), and should therefore react very rapidly with the free radical NO.

It is thus unlikely that the second step of Mechanism 2 would be slow. Mechanism 1 is preferable, based on chemical reasonability. There is yet a fourth mechanism, involving a one-step termolecular process. It is currently believed that this is the correct one.

Let's summarize what we have just learned about reaction mechanisms:

- The steps in a mechanism are numbered from 1 to n. The subscripts on the elementary process rate constants are given the same numbers, with k_n being the forward rate constant and k_{-n} the reverse rate constant for step n. The steps must add to give the overall reaction.
- The mechanism often (but not always) involves at least one slow step. This is the rate-determining step.
- Any fast step preceding the slow step is treated as reversible, and assumed to come to equilibrium.
- To obtain the rate law, we
  - Write the rate law for the elementary rds;
  - Eliminate concentrations of intermediates using rapid pre-equilibrium expressions. This is necessary because the concentrations of intermediates are not usually observable.
- That the reaction proceeds in several steps having different rate constants can cause its rate law to deviate from the overall stoichiometry. However, the rate law tells us which atoms are involved in the reactants of the rate determining step, so is strongly suggestive of mechanism.
- Not all elementary step rate constants need appear in the implied rate law for the mechanism.
- Mechanisms can be disproven, but not proven.

INORGANIC REACTIONS

It should be obvious that particles (ions, atoms, or molecules) of chemically active substances must be brought together before any type of reaction is possible. The most widely used medium for this purpose is water, which may play a multiple role in bringing about chemical reactions between particles. Moreover the reacting particles must be in a dispersed condition before appreciable interaction can take place. Consequently, before entering into a discussion of reactions in aqueous solution, our first task should be an attempt to explain the action of water as a dispersing medium.

The Problem Of Solubility

By definition a solution is a homogeneous mixture of two or more substances, these substances being of molecular dimensions, whose composition may be varied within certain limits. In other words, a solution may be regarded as a molecular dispersion of substances within each other. In aqueous solutions the water is usually considered as the solvent, or medium in which the solute is dissolved. There are, however, conditions where the two components of a simple solution, the solvent and solute, cannot be distinguished from each other, as illustrated by a mixture of water and ethyl alcohol. In a mixture of 75 per cent alcohol and 25 per cent water the alcohol might well be regarded as the solvent and water as the solute. Totally miscible combinations of this type are not common. Usually there is a limit to the solubility of a substance in water, and when this limit is reached, the undissolved solute exists in equilibrium with the dissolved solute, as a saturated solution.

In general the particles encountered in solutions of inorganic substances are derived from solids which have been dissolved in water. There are, however, other mixtures recognized as solutions. For example, gases, such as ammonia and carbon dioxide, or liquids, such as hydrogen cyanide and hydrogen fluoride, can be dissolved in water to form solutions.

The range of solubility of solids in water is quite wide. Potassium nitrate dissolves to the extent of 31.6 g in 100 ml of water at 20°C, whereas silver chloride is soluble only to the extent of 0.0001 g/100 ml of water at 10°C. Why one substance is very soluble and another only slightly soluble may be explained in terms of the attractive forces between the solute particles and solvent molecules. A completely satisfactory analysis of these attractive forces is yet to be made, but sufficient basic facts about solubility are known to enable us to visualize a partial picture of what takes place when a solute dissolves in a solvent.

The Effect of Temperature on Solubility

A rising temperature produces a wide range of effects upon the solubilities of solids in water. Usually the solubilities of substances are increased with an increase in temperature, but there are exceptions to this general rule. Among these exceptions are anhydrous sodium sulfate, whose solubility decreases with rising temperature, and sodium chloride, whose solubility is little affected by temperature change

The solution of a solid is analogous to a melting process. Pure solids are usually crystalline in structure, and the atoms or ions in the crystal lattice are geometrically interspaced so that each ion or atom has the same pattern with respect to neighboring ions or atoms.

The forces which hold these ions or atoms in their relative positions may be quite large. However, when a crystal is heated, the kinetic energy of the particles within the solid is increased.

This increased energy is manifested, at first, by an increased vibration of the particles composing the lattice structure. When the amplitude of vibration becomes sufficiently large, the forces holding the particles together are overcome and the ions or atoms break away from their relatively fixed positions.

In this manner a crystal melts, and the particles of which it was originally composed are free to move throughout the molten substance. In like fashion, when a solid enters into aqueous solution, the solvent particles exert sufficient attraction to overcome the attractive forces holding the solute particles in the crystal lattice. As a general rule, a rise in temperature favors an increase in solubility. The electrostatic forces holding ions or molecules within a crystal are weakened by an increase in the kinetic energy of these particles, thereby enabling the attractive forces of the solvent molecules to be more effective in breaking down the crystal lattice. It is also true that substances with high melting points are more insoluble than those with low melting points. A high melting point indicates a greater intermolecular or interionic attraction. However, so many factors enter into the complicated picture of the solution process that the foregoing explanation is far from complete.

The Role of Water as a Solvent

The electrical dissymmetry of the water molecule, and a diagrammatic representation of the electronic structure of this molecule. The displacement of electric charges within its structure accounts for the polar character of the particle. As a result of the separation of positive and negative charges the water molecule is oriented in an electrostatic field, in a manner similar to that of a small magnet in a magnetic-field. Furthermore the polarity of the H_2O particle enables it to exert an attractive force toward a charged body.

Ionic substances, such as sodium chloride, exist as ions, not only in solution, but also in the crystal. In the process by which a solid of this type enters solution, water molecules, because of their dipole moments, align their negative centers toward the positive ions and their positive ends in the direction of the negative ions at the surface of the crystal.

The dipole forces exerted by the oriented water molecules weaken the interionic attraction within the crystal so that some of the ions are pulled into solution. The removed ions wander away from the crystal and may carry their attached water molecules with them, although there is some doubt as to the advisability of distinguishing between solvent molecules which are attached to an ion and those which remain unattached. If it is assumed that six water molecules are attached to each ion

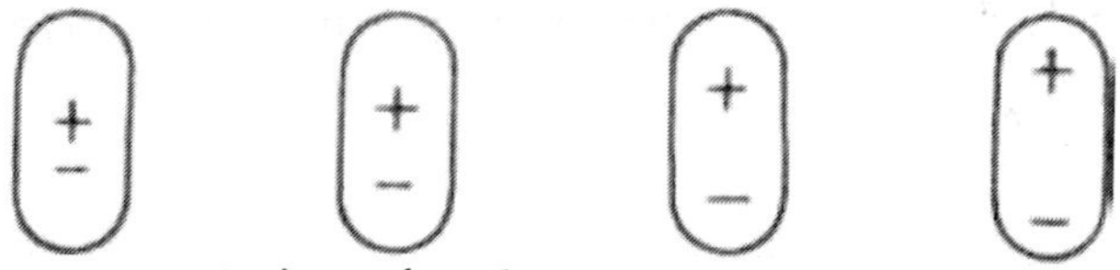

Schematic representation of molecules with increasing degree of polarity.

Ionic compounds are more soluble in solvents with high dielectric constants. This property, sometimes called *specific inductive capacity,* is a measure of the ability of a solvent to modify the mutual interaction of electrified bodies immersed in it, or separated by it. Those solvents which have large dielectric constants also have large separation of charges within their molecules. Consequently the magnitude of the dielectric constant of a solvent is also a measure of its polarity. Furthermore the degree of polarity depends upon the distance between charges within a molecule. Ordinarily, the degree of polarity is designated as the *dipole moment* of the molecule, and the dipole moment is calculated by multiplying the distance between the centers of the two charges by the magnitude of one of the charges. Thus

$$\mu = de$$

where μ denotes the dipole moment. The polar molecules with different polarities resulting from differences in the extent of the separation of the intramolecular charges. The dipole moment is always of the order of 10^{-18} esu. This magnitude is to be anticipated since the unit electronic charge is 4.8×10^{-10} esu and the diameter of most molecules is of the order of 10^{-8} cm. A number of liquid compounds according to their decreasing dipole moments.

Dipole Moments of a Number of Liquids

Liquid compound	Dipole moment, esu
C6H5NO2(nitrobenzene)	4.18×10^{-18}
H2O (water)	1.84×10^{-18}
C2H5OH (ethyl alcohol)	1.70×10^{-18}
C6H5OH (phenol)	1.70×10^{-18}
NH3 (liquid ammonia)	1.46×10^{-18}
CHCl3 (chloroform)	1.15×10^{-18}
HBr (liquid hydrogen bromide)	0.79×10^{-18}
C6H6 (benzene)	0
CCl4 (carbon tetrachloride)	0

The dipoles in water are attracted to each other by fairly strong forces called *hydrogen bonds.* Consequently, for a substance to be soluble in water, it must be capable of weakening the hydrogen bonding in order that the solute particles may penetrate into the intermolecular spaces of the solvent. When water molecules are subjected to the electrostatic field of an ion, there is a tendency for the dipole to be aligned, or rotated, in the direction of the field. Any alignment, or rotation, results in a slight displacement of the oppositely charged parts of each molecule and a consequent loss of potential energy by the molecules of the solvent. In other words, the dipoles of water are attracted by the electrostatic charges of ions, resulting in a weakening of hydrogen bonding in the solvent and a penetration by the ion into the intermolecular spaces of the liquid. On the other hand bonding forces between polar molecules of water prevent the entrance of nonpolar molecules into the intermolecular spaces of the solvent. It might be said that nonpolar molecules do not enter into solution in a polar solvent because of a squeezing-out effect. Thus water dissolves an ionic solid, such as sodium chloride, but will not dissolve paraffin, which is a nonionic and a nonpolar solid. For similar reasons water dissolves ammonia, both substances being strongly polar, but will not dissolve a nonpolar gas such as methane.

Chemical Structure and Solubility

"Like dissolves like" is a familiar rule in organic chemistry and is also applicable to inorganic chemistry. This rule refers to similarity in relation not only to dipole moment but also to chemical composition. It is generally true that a substance tends to dissolve in a solvent which is chemically related in composition to the substance. For example, paraffin readily dissolves in gasoline, or water is good solvent of cane sugar. Gasoline and paraffin are both hydrocarbons — compounds containing only carbon and hydrogen. Cane sugar is a carbohydrate and, as the name "carbohydrate " implies, is a composition of carbon, hydrogen, and oxygen in which the ratio of hydrogen to oxygen is the same as in water.

The explanation of the process of solution becomes more complex when the solvent is a liquid whose molecules are associated as a result of hydrogen bonding. Among solvents of this type are water, liquid ammonia, and liquid hydrogen fluoride.

The hydrogen bond is produced where the hydrogen atom serves as a bridge between two strongly electronegative atoms. Consequently molecules which contain hydrogen attached to the more electronegative atoms tend to group into clusters which are held together by hydrogen bridges.

The tetrahedral arrangement of water molecules in ice is a result of hydrogen bonding, and this tetrahedral structure persists, to some extent, after ice is converted into a liquid. Therefore the composition of water is not that of simple H_2O molecules but that of multiple molecules held together somewhat loosely by hydrogen bonds.

Water, liquid ammonia, liquid hydrogen fluoride, and solvents of this type, because of their associated structures, will not necessarily dissolve polar molecules unless the solute molecules can form hydrogen bonds with the solvent molecules.

For example, nitrobenzene, $C_6H_5NO_2$, has a dipole moment of 4.18 ◎× 10^{-18} esu, which is large as compared with the dipole moment of water, but nitrobenzene molecules are practically insoluble in water because they cannot form hydrogen linkages with water molecules. On the other hand phenol, C_6H_5OH, is more than forty times as soluble in water as nitrobenzene. Phenol, which is similar to nitrobenzene in that it is a monosubstituted benzene product, has a dipole moment of 1.7×10^{-18} esu and should be less soluble than nitrobenzene if relative solubility were entirely dependent upon the magnitude of the dipole moment. This unexpected solubility must be due to the formation of hydrogen bonds between the hydroxyl groups of the phenol molecules and the oxygen atoms in the water molecules.

Another complex form of solubility is presented by organic molecules whose structures are such that they are partly polar and partly nonpolar. A molecule of this type is exemplified by heptyl alcohol, $CH_3CH_2CH_2CH_2CH_2CH_2CH_2OH$, which is composed of a hydrocarbon chain with a hydroxyl group attached to one end. The hydroxyl end of this molecule is soluble, because it penetrates the structure of water to form hydrogen bonds with the solvent. The hydrocarbon end of the molecule is nonpolar and cannot form hydrogen bonds and is therefore insoluble in water. Molecules of heptyl alcohol, when placed in water, are pushed to the surface, where they orient themselves so that the soluble ends are in the water and the nonpolar ends protrude above the surface.

It is of interest that hydrocarbon molecules with soluble polar end groups such as $-CH_2OH$ or $--COOH$ may form unimolecular (a depth of one molecule) films upon the surface of water. Films of this type also occur at any interface of water and its container, and as a result the surface tension of water and its "wetting" properties may undergo decided changes. Wetting agents of this general type have wide commercial applications.

The size, charge, and electronic structure of each ion in the crystal lattice of an ionic compound affect the solubility of the particular electrolyte. Although certain generalizations regarding the influences of these factors are somewhat obvious, it is impossible to arrive at any hard and fast rules concerning these structural variations. Large ions should pack less tightly in a lattice than small ions of similar charges, because of the greater distances between the effective charges borne by the larger particles. The radii of the sodium and fluoride ions are listed as 0.95 and 1.36 A, respectively, whereas the radii of the cesium and iodide ions are given as 1.69 and 2.16 A. Consequently it may be assumed that the lattice energy of cesium iodide is less than that of sodium fluoride because of the differences in the radii of the ions which make up the two compounds. The values given for certain lattice energie verify the foregoing assumption. According to this table the lattice energy of sodium fluoride is

213.4 kcal, and that of cesium iodide is 139.1 kcal. Furthermore solubility data for the two compounds indicate that cesium iodide is approximately ten times as soluble in water as is sodium fluoride.

High ionic charges should decrease the solubility of an electrolyte. The effect of increased ionic charge upon solubility is illustrated by the compounds magnesium oxide and barium oxide.

Since the ions in these two substances are doubly charged, the particles should pack more tightly in the crystal lattice than might be expected in the case of monovalent ions.

The radii of the magnesium, barium, and oxygen ions are listed as 0.65, 1.35, and 1.40 A, respectively, and these values would indicate that the ions of magnesium oxide should be held more firmly in its lattice than are the ions in barium oxide. Such a prediction turns out to be true, inasmuch as the lattice energy of magnesium oxide is given as 940 kcal as compared with a value of 747 kcal for barium oxide. A comparison of the solubilities of the two compounds shows that barium oxide is over 2,000 times more soluble than magnesium oxide.

From the examples given in the preceding paragraphs it might appear that the tightness with which ions are held in a crystal lattice should always involve the simple relationship of an attractive force operating between two charged bodies. In other words, one might conclude that the strength of the coulombic bond between two ions is always proportional to the product of the two charges and inversely proportional to the square of the distance between the effective centers of the two ions.

Such a conclusion would be erroneous, as is indicated by the relative solubilities of magnesium sulfate and barium sulfate. The radii of the doubly charged magnesium, barium, and sulfate ions are 0.65, 1.35, and 3.0 A, respectively, and from these values it might be wrongly deduced that both compounds are insoluble and that barium sulfate is more soluble than magnesium sulfate.

Nevertheless it so happens that magnesium sulfate is quite soluble and barium sulfate is fairly insoluble. The explanation for this apparent anomaly lies in the large difference between the radii of the magnesium and sulfate ions. On the basis of coulombic attraction alone, magnesium sulfate should be quite insoluble; however, the small magnesium ions cannot fill up the large spaces between the sulfate ions in the crystal lattice.

In order to attain effective packing, each magnesium ion is surrounded by a number of water molecules in such wise that the resulting hydrated particle is sufficiently large to fill its allotted space in the lattice. As a result of this hydration the force operating between the hydrated magnesium ion and the sulfate ion is so weak that the ions are easily separated in producing an aqueous solution.

The relation between solubility and binding forces in a crystal lattice is far from simple. In a crystal produced by the deposition of ions there frequently appears a tendency for the electrovalencies of the constituent particles to acquire a covalent character. This tendency is designated as *polarization* and usually depends upon the extent to which the anion may be deformed and upon the polarizing power of the cation. In other words, a small positive ion may distort a large negative ion, and in the process an electron may be pulled away from the anion to form an electron pair or covalent linkage. The effect of polarization may be illustrated by the structures of the silver halides. The crystals of AgF, AgCl, and AgBr have a sodium chloride crystal structure, but as the anion in these compounds increases in size, the polarizing effect of the small Ag^+ ion becomes stronger. In fact the bonding in AgI has very little ionic character, and the crystal has a structure that is characteristic of certain covalent compounds, usually designated as the zinc blende, ZnS, structure. Spectroscopic data indicate that silver iodide is composed of atoms and not of ions.

REACTION RATES

Chemical kinetics is the branch of chemistry which addresses the question: "how fast do reactions go?" Chemistry can be thought of, at the simplest level, as the science that concerns itself with making new substances from other substances. Or, one could say, chemistry is taking molecules apart and putting the atoms and fragments back together to form new molecules. (OK, so once in a while one uses atoms or gets atoms, but that doesn't change the argument.) All of this is to say that chemical reactions are the core of chemistry.

If Chemistry is making new substances out of old substances (i.e., chemical reactions), then there are two basic questions that must be answered:

- Does the reaction want to go? This is the subject of chemical thermodynamics.
- If the reaction wants to go, how fast will it go? This is the subject of chemical kinetics.

Here are some examples. Consider the reaction,

$$2\ H_2(g) + O_2(g) \rightarrow 2\ H_2O(l).$$

We can calculate $\Delta_r G^o$ for this reaction from tables of free energies of formation (actually this one is just twice the free energy of formation of liquid water). We find that $\Delta_r G^o$ for this reaction is very large and negative, which means that the reaction wants to go very strongly. A more scientific way to say this would be to say that the equilibrium constant for this reaction is very

very large. However, we can mix hydrogen gas and oxygen gas together in a bulb or other container, even in their correct stoichiometric proportions, and they will stay there for centuries, perhaps even forever, without reacting. (If we drop in a catalyst - say a tiny piece of platinum - or introduce a spark, or even illuminate the mixture with sufficiently high frequency uv light, or compress and heat the mixture, the mixture will explode.) The problem is not that the reactants do not want to form the products, they do, but they cannot find a "pathway" to get from reactants to products.

Another example: consider the reaction,

$$C(\textit{diamond}) \rightarrow C(\textit{graphite}).$$

If you calculate $\Delta_r G^o$ for this reaction from data in the tables of thermodynamic properties you will find once again that it is negative (not very large, but still negative). This result tells us that diamonds are thermodynamically unstable. Yet diamonds are highly regarded as gem stones ("diamonds are forever") and are considered by some financial advisors as a good long-term investment hedge against inflation. On the other hand, if you were to vaporize a diamond in a furnace, under an inert atmosphere, and then condense the vapour, the carbon would come back as graphite and not as diamond.

How can all these things be?

The answer is that thermodynamics is not the whole story in chemistry. Not only do we have to know whether a reaction is thermodynamically favored, we also have to know whether the reaction can or will proceed at a finite rate. The study of the rate of reactions is called chemical kinetics.

The study of chemical kinetics requires new definitions, new types of experimental data, and new theories and equations to organize the data. We begin with the definition of *reaction rate*.

Reaction Rates

Consider the reaction,

$$2\ NO(g) + O_2(g) \rightarrow 2\ NO_2(g).$$

We can specify the rate of this reaction by telling the rate of change of the partial pressures of one the gases. However, it is convenient to convert these pressures into concentrations, so we will write our rates and rate equations in terms of concentrations, where square brackets, [ ], mean concentration in mol/L.

We might try to write the rate variously as, $\frac{d[NO_2]}{dt}$,

or as

$-\frac{d[O_2]}{dt}$,

but these are not the same because each molecule of O_2 gives two molecules of NO_2. To arrive at an unambiguous definition of reaction rate we define the "reaction velocity," v, as

$$v = \frac{1}{2}\frac{d[NO_2]}{dt} = -\frac{d[O_2]}{dt} = -\frac{1}{2}\frac{d[NO]}{dt}.$$

This is unambiguous. The negative sign tells us that that species is being consumed and the fractions take care of the stoichiometry. Any one of the three derivatives can be used to define the rate of the reaction.

For a general reaction,

$$aA + bB \rightarrow cC + dD,$$

the reaction velocity can be written in a number of different but equivalent ways,

$$v = \frac{1}{a}\frac{d[A]}{dt} = -\frac{1}{b}\frac{d[B]}{dt} = -\frac{1}{c}\frac{d[C]}{dt} = \frac{1}{d}\frac{s[D]}{dt}.$$

As in our previous example, the negative signs account for material that is being consumed in the reaction and the positive signs account for material that is being formed in the reaction. The stoichiometry is preserved by dividing the rate of change of concentration of each substance by its stoichiometric coefficient.

Rate Laws

A rate law is an equation that tells us how fast the reaction proceeds and how the reaction rate depends on the concentrations of the chemical species involved. A rate law is an equation of the form,

$$v = f([A],[B],...[E]).$$

Equation 4 is gives us a first order differential equation in t because the reaction velocity is related to a time-derivative of one of the concentrations.

The rate law may contain substances which are not in the balanced reaction and may not contain some things that are in the balanced equation (even on the reactant side).

Usually rate laws take the form,

$$v - k([A]^x [B]^y [E])^z ...,$$

where x, y, z, are small whole numbers or simple fractions and k is called the "rate constant." The sum of $x + y + z + ...$ is called the "order" of the reaction.

Common Types of Rate Laws

- First Order Reactions

  In a first order reaction the rate is proportional to the concentration of one of the reactants. That is,

v = rate = k[B],

where B is a reactant. If we have a reaction which is known to be first order in B, such as

B + other reactants → products,

we would write the rate law as,

$$-\frac{d[B]}{dt} = [B].$$

The constant, k, in this rate equation is the first order rate constant.

- Second Order Reactions

  In a second order reaction the rate is proportional to concentration squared.For example, possible second order rate laws might be written as

  Rate = $k[B]^2$

  or as

  Rate = k[A][B].

  That is, the rate might be proportional to the square of the concentration of one of the reactants, or it might be proportional to the product of two different concentrations.

- Third Order Reactions

  There are several different ways to write a rate law for a third order reaction. One might have cases where

  Rate = $k[A]^3$,

  or

  Rate = $k[A]^2[B]$,

  or

  Rate = k[A][B][C],

  and so on.

We will see later that there are other, more "interesting" rate laws in nature, but a large fraction of rate laws will fit in one of the above categories.

Integrated Forms of Rate Laws

In order to understand how the concentrations of the species in a chemical reaction change with time it is necessary to integrate the rate law (which is given as the time-derivative of one of the concentrations) to find out how the concentrations change over time.

- First Order Reactions

  Suppose we have a first order reaction of the form,

  B +.... → products.

Then we can write the rate law and integrate it as follows (recall that the derivative is negative because the concentration of the reactant, B, is decreasing):

$$\frac{d[B]}{dt} + -k[B]$$

$$\frac{d[B]}{B} + -k[dt]$$

$$\text{In}\frac{[B]}{[B]_\circ} + -kt$$

$$[B]=[B]_\circ e^{-kt}$$

The first order rate law is a very important rate law, radioactive decay and many chemical reactions follow this rate law and some of the language of kinetics comes from this law.

The form of Equation 14d is called an "exponential decay." This form appears in many places in nature. One of its consequences is that it gives rise to a concept called "half-life."

Half-life

The half-life, usually symbolized by $t_{1/2}$, is the time required for [B] to drop from its initial value $[B]_o$ to $[B]_o/2$.

Using the integrated form of the first order rate law we find that

$$\frac{[B]_\circ}{2}=[B]_\circ e^{-kt_{1/2}},$$

or

$$\frac{1}{2}=e^{-kt_{1/2}}$$

Taking the logarithm of both sides gives,

$$\text{In}\frac{1}{2}=-kt_{1/2},$$

$$\text{In}2=kt_{1/2}$$

or

$$t_{1/2}=\frac{\text{In}2}{K}.$$

(You can also write

$$[B]=[B]_\circ 2^{-\frac{t}{t_{1/2}}}=\frac{[B]_\circ}{2^{\frac{t}{t_{1/2}}}},$$

which may actually give a little more insight into what is meant by half-life. This equation demonstrates clearly that the concentration drops by a factor of two for every $t_{1/2}$ increment in time.)

For first order processes it is common to define a "relaxation time." τ, by

$$\tau = \frac{1}{k},$$

so that one can write the integrated form of the rate law as

$$[B] = [B]_o \, e^{-t/\tau}.$$

τ is the time required for [B] to drop from $[B]_o$ to $[B]_o/e$. Sometimes τ is called the "one over *e*" time. Although the half-life is almost always used to describe the decay rate of radioactive elements, it is common for chemists to talk about the rate of first order processes in chemistry in terms of the relaxation time.

EXPERIMENTAL DETERMINATION OF THE RATE LAW THE INITIAL RATE METHOD

When applicable, the method of initial rates is a simple and clean way to obtain the rate law. The initial rate of a reaction is the instantaneous rate at t=0, when reactants are first mixed together and begin to react. In a plot of reactant concentration versus time, it is given by the slope of the tangent at t = 0.

The slope of this type of plot is always steepest at the beginning; the initial rate is thus the fastest rate that we observe during the time course of the reaction process.

One major advantage of the initial rate method is that the reverse rate is zero, provided only reactants and NO products are initially present. Thus only the forward rate contributes to the initial rate, and complications resulting from the reverse reaction are avoided. The initial rate law depends only on concentrations of reactants. The method of initial rates is best illustrated by example.

Example: The initial rate was measured at several combinations of initial reactant concentration for the following reaction:

$$A + 2B + 3C \rightarrow D + E$$

The rate law is expected to have the following general form:

$$-d[A]/dt = k[A]^x[B]^y[C]^z$$

The following data were obtained. Use the data to determine the values of x, y and z.

Run Number	$[A]_o$, M	$[B]_o$, M	$[C]_o$, M	Initial Rate, M/s
1	0.05	0.10	0.10	6.00×10^{-3}
2	0.05	0.15	0.10	13.5×10^{-3}
3	0.05	0.15	0.05	13.5×10^{-3}
4	0.10	0.10	0.05	12.0×10^{-3}

Solution: The key to the initial rate method is to examine the dependence of rate on one reactant at a time, using data obtained when the other reactants do not vary in initial concentration. In the first two data entries above, the initial concentrations of both A and C are the same in both runs; only the initial concentration of B is changed. These two data items can therefore be used to determine the order of the reaction in B. We see that when the initial concentration of B is increased from 0.10 to 0.15, an increase of a factor of 1.5, the rate increases from 6 to 13.5×10^{-3} moles/L-s, which is a factor of 2.25. We translate this information into equation form:

$$\text{Rate 2/Rate 1} = [B]_2^y/[B]_1^y = ([B]_2/[B]_1)^y$$

$$2.25 = (1.5)^y$$

We recognize that y must be 2. The reaction is therefore second order in B, and the rate law thus far is

$$-d[A]/dt = k[A]^x[C]^z[B]^2$$

In runs 2 and 3, the initial concentrations of A and B are the same, while that of C is changed. We see from these data that even though the concentration of C is only half as large in run 3 as in run 2, the rate is unaffected. We conclude that the rate does not depend on [C], hence that $z = 0$. The rate law to this point is

$$-d[A]/dt = k[A]^x[B]^2$$

There are no two kinetics runs in which only [A] is varied. However, runs 1 and 4 can be used to obtain the order in A because we have just determined that rate does not depend on [C]. Runs 1 and 4 indicate that the rate doubles when the initial concentration of A doubles. The order in A is thus 1, and the final rate law is

$$-d[A]/dt = k[A][B]^2$$

Once the rate law is known, the rate constant may be calculated from any one of the individual runs. Choosing the first run,

$$k = \text{Rate}/[A][B]^2 = 6.00 \times 10^{-3}/(0.05)(0.10)^2 = 12.0 \text{ L}^2/\text{mole}^2\text{s}$$

As this example shows, the initial rate method, when it is applicable, provides a clean, simple path to the rate law. The italicized qualifier is important, however; in practice, the method of initial rates is only rarely used, for several reasons. In most kinetics studies, the chemist examines the manner in which the concentration of a selected reagent varies with time. The resulting data are presented either in a table, or in a plot of concentration versus time. The data obtained at the beginning of the reaction, where the rate is largest and the time interval is smallest, is usually less reliable than that obtained at later times, when the rate is less. It is undesirable to base an analysis on one's least reliable data. Second, and more important, the method of initial rates uses only a small fraction of the concentration time data; the rest is wasted.

Third, to apply the initial rate method as was done in the example above requires that full concentration-time profiles be collected for each run, in order to extract the slope of the concentration-time plot at t=0. Collection of this much data requires a lot of experimental effort. Most of the resulting data is then ignored if the initial rate method is used. For these reasons and others, chemists have developed other methods for determining rate laws from data, which make use of all of the data collected.

Integration of the Rate Law

It is possible to convert the differential rate law into an integrated form, in which concentration is expressed directly as a function of time. The concentration-time data obtained during a kinetics run in the lab may then be matched against the function. If the data conform to the function, it may be concluded that the reaction follows the corresponding differential rate law. Integration is very easy for rate laws of the form,

$$-d[A]/dt = k[A]^n$$

We will examine the first and second-order cases, with n = 1 and 2 respectively. The first order rate law may be rearranged to put the concentration terms on one side and time on the other side of the equals sign:

$$-d[A]/[A] = kdt$$

Both sides of the equation are then integrated over the limits, $[A] = [A]_o$ at t=0 and $[A] = [A]_t$ at some later time t:

$$\int_{[A]0}^{[A]t} d[A]/[A] = k\int_0^t dt$$

The integral of dx/x is ln x, and the integral of dx is x. Evaluating over the limits gives:

$$\ln([A]_t/[A]_o) = -k\,t$$

This is the so-called integrated first order rate law. For a reaction following simple first order kinetics, a plot of ln $[A]_t$ versus t is linear, with slope -k. Thus if concentration-time data for a reaction are plotted according to equation, and linearity is obtained, it may be concluded that the reaction follows a simple first-order differential rate law. An alternative expression is obtained by exponentiating both sides:

$$[A]_t = [A]_o e^{-kt}$$

Here e = 2.718... is the base of natural logarithms. Substituting the reciprocal of the time constant, t, for the rate constant gives

$$[A]_t = [A]_o e^{-t/t}$$

Thus when an interval of time equal to the time constant has elapsed, the concentration of A will have decreased to 1/e (about 1/3) of its initial value.

With the integrated first-order rate law in hand, we are in a position to consider a quantity called the half-life, $t_{1/2}$, which is similar but not identical to the time constant. The half-life for a process is defined as the time required for the concentration of the limiting reactant to decrease to 1/2 of its initial value. An expression for the half life is easily obtained from by realizing that $[A]_t = [A]_o/2$ at $t = t_{1/2}$. Making these substitutions leads to the following expression for the half-life of a first-order reaction:

$$t_{1/2} = \ln 2/k = 0.693/k$$

For n = 2, corresponding to a second-order process, rearrangement leads:

$$: -d[A]/[A]^2 = k\ dt$$

Integration of both sides of this equation results:

$$1[A] - 1/[A]_o = k\ t$$

For a reaction with a simple second order rate law, a plot of concentration-time data as 1/[A] versus t is linear, with slope k. Thus a reaction for which the concentration-time data produce a linear 1/[A] versus t plot may be concluded to follow a second-order rate law.

Example: Plot the concentration-time data in Example.according to equations to determine whether the hydrolysis of *t*-butyl bromide obeys a simple first- or second-order rate law.

Solution: The concentration-time data from Example.are reproduced below. Columns have been added for ln [*t*-BuBr] and 1/[*t*-BuBr]

time, s	**[t-BuBr], moles/L**	**ln [t-BuBr]**	**1/[tBuBr]**
0	0.100	–2.303	10
10	0.0876	–2.435	11.42
20	0.0768	–2.567	13.02
30	0.0672	–2.700	14.88
40	0.059	–2.830	16.95
50	0.0517	–2.962	19.34
60	0.0453	–3.094	22.08
80	0.0348	–3.358	28.74
100	0.0267	–3.623	37.45
120	0.0205	–3.887	48.78
180	0.0093	–4.678	107.53
240	0.0042	–5.473	238.10

The linearity of the first plot indicates that the reaction is first-order in *t*-BuBr. The rate constant, obtained from the slope is 1.32×10^{-2} s^{-1} (recall that for a first order reaction, the units of the rate constant must be $time^{-1}$, resulting from the division of rate in conc/time by concentration).

Example: One product resulting from fission of ^{235}U in nuclear reactors is ^{137}Cs, a radioactive nucleus with a time constant of 43.3 years. How much time will it take for the radioactivity level of this waste to fall to 1% of its initial value, which is considered a safe level?

Solution: This problem is readily solved using equation :

$$\ln [A]/[A]_o = -kt = -t/t$$

$$t = -\ln(0.01)t = -\ln(0.01)(43.3) = 200 \text{ years}$$

The waste must be safely stored for 200 years before it will be safe.

5

Structure and Properties of Polymers

Having established the basic principles of classification in polymer chemistry, we will now turn our attention to individual polymers and consider a little about their physical and chemical properties. Most of the examples which follow are of commercial importance, though it is their properties that are emphasised rather than the complete chemistry of their industrial manufacture.

In terms of the classification systems just referred to, it is found that the poly-mers of major commercial importance tend to be the thermoplastics or addi-tion polymers rather than the thermosets, and these will be dealt with first.

THE BASICS: POLYMER DEFINITION AND PROPERTIES

The simplest definition of a polymer is a useful chemical made of many repeating units. A polymer can be a three dimensional network (think of the repeating units linked together left and right, front and back, up and down) or two-dimensional network (think of the repeating units linked together left, right, up, and down in a sheet) or a one-dimensional network (think of the repeating units linked left and right in a chain).

Each repeating unit is the "-mer" or basic unit with "poly-mer" meaning many repeating units. Repeating units are often made of carbon and hydrogen and sometimes oxygen, nitrogen, sulfur, chlorine, fluorine, phosphorous, and silicon. To make the chain, many links or "-mers" are chemically hooked or polymerized together. Linking countless strips of construction paper together to make paper garlands or hooking together hundreds of paper clips to form chains, or stringing beads helps visualize polymers.

Polymers occur in nature and can be made to serve specific needs. Manufactured polymers can be three-dimensional networks that do not melt once formed. Such networks are called THERMOSET polymers. Epoxy resins used in two-part adhesives are thermoset plastics. Manufactured polymers can also be one-dimensional chains that can be melted. These chains are THERMOPLASTIC polymers and are also called LINEAR polymers. Plastic bottles, films, cups, and fibers are thermoplastic plastics.

Polymers abound in nature. The ultimate natural polymers are the deoxyribonucleic acid (DNA) and ribonucleic acid (RNA) that define life. Spider silk, hair, and horn are protein polymers. Starch can be a polymer as is cellulose in wood. Rubber tree latex and cellulose have been used as raw material to make manufactured polymeric rubber and plastics.

The first synthetic manufactured plastic was Bakelite, created in 1909 for telephone casing and electrical components. The first manufactured polymeric fiber was Rayon, from cellulose, in 1910. Nylon was invented in 1935 while pursuing a synthetic spider silk.

The Structure of Polymers

Many common classes of polymers are composed of hydrocarbons, compounds of carbon and hydrogen. These polymers are specifically made of carbon atoms bonded together, one to the next, into long chains that are called the backbone of the polymer.

Because of the nature of carbon, one or more other atoms can be attached to each carbon atom in the backbone. There are polymers that contain only carbon and hydrogen atoms. Polyethylene, polypropylene, polybutylene, polystyrene and polymethylpentene are examples of these. Polyvinyl chloride (PVC) has chlorine attached to the all-carbon backbone. Teflon has fluorine attached to the all-carbon backbone.

Other common manufactured polymers have backbones that include elements other than carbon. Nylons contain nitrogen atoms in the repeat unit backbone. Polyesters and polycarbonates contain oxygen in the backbone. There are also some polymers that, instead of having a carbon backbone, have a silicon or phosphorous backbone. These are considered inorganic polymers. One of the more famous silicon-based polymers is Silly Putty®.

Molecular Arrangement of Polymers

Think of how spaghetti noodles look on a plate. These are similar to how linear polymers can be arranged if they lack specific order, or are amorphous. Controlling the polymerization process and quenching molten polymers can result in amorphous organization. An amorphous arrangement of molecules has no long-range order or form in which the polymer chains arrange themselves. Amorphous polymers are generally transparent. This is an important characteristic for many applications such as food wrap, plastic windows, headlight lenses and contact lenses.

Obviously not all polymers are transparent. The polymer chains in objects that are translucent and opaque may be in a crystalline arrangement. By definition, a crystalline arrangement has atoms, ions, or in this case, molecules arranged in distinct patterns.

You generally think of crystalline structures in table salt and gemstones, but they can occur in plastics. Just as quenching can produce amorphous arrangements, processing can control the degree of crystallinity for those polymers that are able to crystallize. Some polymers are designed to never be able to crystallize.

Others are designed to be able to be crystallized. The higher the degree of crystallinity, generally, the less light can pass through the polymer. Therefore, the degree of translucence or opaqueness of the polymer can be directly affected by its crystallinity. Crystallinity creates benefits in strength, stiffness, chemical resistance, and stability.

Scientists and engineers are always producing more useful materials by manipulating the molecular structure that affects the final polymer produced. Manufacturers and processors introduce various fillers, reinforcements and additives into the base polymers, expanding product possibilities.

CHARACTERISTICS OF POLYMERS

The majority of manufactured polymers are thermoplastic, meaning that once the polymer is formed it can be heated and reformed over and over again. This property allows for easy processing and facilitates recycling. The other group, the thermosets, cannot be remelted. Once these polymers are formed, reheating will cause the material to ultimately degrade, but not melt.

Every polymer has very distinct characteristics, but most polymers have the following general attributes:

1. Polymers can be very resistant to chemicals. Consider all the cleaning fluids in your home that are packaged in plastic. Reading the warning labels that describe what happens when the chemical comes in contact with skin or eyes or is ingested will emphasize the need for chemical resistance in the plastic packaging. While solvents easily dissolve some plastics, other plastics provide safe, non-breakable packages for aggressive solvents.
2. Polymers can be both thermal and electrical insulators. A walk through your house will reinforce this concept, as you consider all the appliances, cords, electrical outlets and wiring that are made or covered with polymeric materials. Thermal resistance is evident in the kitchen with

pot and pan handles made of polymers, the coffee pot handles, the foam core of refrigerators and freezers, insulated cups, coolers, and microwave cookware. The thermal underwear that many skiers wear is made of polypropylene and the fiberfill in winter jackets is acrylic and polyester.

3. Generally, polymers are very light in weight with significant degrees of strength. Consider the range of applications, from toys to the frame structure of space stations, or from delicate nylon fiber in pantyhose to Kevlar, which is used in bulletproof vests. Some polymers float in water while others sink. But, compared to the density of stone, concrete, steel, copper, or aluminum, all plastics are lightweight materials.
4. Polymers can be processed in various ways. Extrusion produces thin fibers or heavy pipes or films or food bottles. Injection molding can produce very intricate parts or large car body panels. Plastics can be molded into drums or be mixed with solvents to become adhesives or paints. Elastomers and some plastics stretch and are very flexible. Some plastics are stretched in processing to hold their shape, such as soft drink bottles. Other polymers can be foamed like polystyrene (Styrofoam™), polyurethane and polyethylene.
5. Polymers are materials with a seemingly limitless range of characteristics and colors. Polymers have many inherent properties that can be further enhanced by a wide range of additives to broaden their uses and applications. Polymers can be made to mimic cotton, silk, and wool fibers; porcelain and marble; and aluminum and zinc. Polymers can also make possible products that do not readily come from the natural world, such as clear sheets and flexible films.
6. Polymers are usually made of petroleum, but not always. Many polymers are made of repeat units derived from natural gas or coal or crude oil. But building block repeat units can sometimes be made from renewable materials such as polylactic acid from corn or cellulosics from cotton linters. Some plastics have always been made from renewable materials such as cellulose acetate used for screwdriver handles and gift ribbon. When the building blocks can be made more economically from renewable materials than from fossil fuels, either old plastics find new raw materials or new plastics are introduced.
7. Polymers can be used to make items that have no alternatives from other materials. Polymers can be made into clear, waterproof films. PVC is used to make medical tubing and blood bags that extend the shelf life of blood and blood products. PVC safely delivers flammable oxygen in non-burning flexible tubing. And anti-thrombogenic material,

such as heparin, can be incorporated into flexible PVC catheters for open heart surgery, dialysis, and blood collection. Many medical devices rely on polymers to permit effective functioning.

Solid Waste Management

In addressing all the superior attributes of polymers, some of the challenges associated with the materials. Most plastics deteriorate in full sunlight, but never decompose completely when buried in landfills. However, other materials such as glass, paper, or aluminum do not readily decompose in landfills either. Some bioplastics do decompose to carbon dioxide and water, however, in specially designed food waste commercial composting facilities ONLY. They do not biodegrade under other circumstances.

For 2005 the EPA characterization of municipal solid waste before recycling for the United States showed plastics made up 11.8 percent of our trash by weight compared to paper that constituted 34.2 percent. Glass and metals made up 12.8 percent by weight. And yard trimmings constituted 13.1 percent of municipal solid waste by weight.

Food waste made up 11.9 percent of municipal solid waste. The characteristics that make polymers so attractive and useful, lightweight and almost limitless physical forms of many polymers designed to deliver specific appearance and functionality, make post-consumer recycling challenging. When enough used plastic items can be gathered together, companies develop technology to recycle those used plastics.

The recycling rate for all plastics is not as high as any would want. But, the recycling rate for the 1,170,000,000 pounds of polyester bottles, 23.1%, recycled in 2005 and the 953,000,000 pounds of high density polyethylene bottles, 28.8%, recycled in 2005 show that when critical mass of defined material is available, recycling can be a commercial success.

Applications for recycled plastics are growing every day. Recycled plastics can be blended with virgin plastic (plastic that has not been processed before) without sacrificing properties in many applications. Recycled plastics are used to make polymeric timbers for use in picnic tables, fences and outdoor playgrounds, thus providing low maintenance, no splinters products and saving natural lumber. Plastic from soft drink and water bottles can be spun into fiber for the production of carpet or made into new food bottles. Closed loop recycling does occur, but sometimes the most valuable use for a recycled plastic is into an application different than the original use.

An option for plastics that are not recycled, especially those that are soiled, such as used food wrap or diapers, can be a waste-to-energy system (WTE). In 2005, 13.6% of US municipal solid waste was processed in WTE systems· When localities decide to use waste-to-energy systems to manage solid waste, plastics can be a useful component.

The controlled combustion of polymers produces heat energy. The heat energy produced by the burning plastic municipal waste not only can be converted to electrical energy but also helps burn the wet trash that is present.

Paper also produces heat when burned, but not as much as do plastics. On the other hand, glass, aluminum and other metals do not release any energy when burned.

To better understand the incineration process, consider the smoke coming off a burning item. If one were to ignite the smoke with a lit propane torch, one would observe that the smoke disappears. This exercise illustrates that the by-products of incomplete burning are still flammable. Proper incineration burns the material and the by-products of the initial burning and also takes care of air and solid emissions to insure public safety.

Some plastics can be composted either because of special additives or because of the construction of the polymers. Compostable plastics frequently require more intense conditions to decompose than are available in backyard compost piles. Commercial composters are suggested for compostable plastics. In 20051, composting processed 8.4% of US municipal solid waste.

Plastics can also be safely land filled, although the valuable energy resource of the plastics would then be lost for recycling or energy capture. In 20051, 54.3% of US municipal solid waste was land filled. Plastics are used to line landfills so that leachate is captured and groundwater is not polluted. Non-degrading plastics help stabilize the ground so that after the landfill is closed, the land can be stable enough for useful futures.

Polymers affect every day of our life. These materials have so many varied characteristics and applications that their usefulness can only be measured by our imagination. Polymers are the materials of past, present and future generations.

CLASSIFICATION OF POLYMERS

There are a number of methods of classifying polymers. One is to adopt the approach of using their response to thermal treatment and to divide them into thermoplastics and thermosets. Thermoplastics are polymers which melt when heated and resolidify when cooled, while thermosets are those which do not melt when heated but, at sufficiently high temperatures, decompose irreversibly.

This system has the benefit that there is a useful chemical distinction between the two groups. Thermoplastics comprise essentially linear or lightly branched polymer molecules, while thermosets are substantially crosslinked materials, consisting of an extensive three-dimensional network of covalent chemical bonding.

Another classification system, first suggested by Carothers in 1929, is based on the nature of the chemical reactions employed in the polymerisation. Here the two major groups are the condensation and the addition polymers. Condensation polymers are those prepared from monomers where reaction is accompanied by the loss of a small molecule, usually of water, for example polyesters which are formed by the condensation shown in Reaction.

$$n\,HO{-}R{-}OH + n\,HOOC{-}R^1{-}COOH \longrightarrow$$

$$HO[{-}R{-}COO{-}R^1{-}COO{-}]_nH + (n-1)H_2O$$

By contrast, addition polymers are those formed by the addition reaction of an unsaturated monomer, such as takes place in the polymerisation of vinyl chloride.

$$n\,CH_2{=}CHCl \longrightarrow [{-}CH_2{-}CHCl{-}]_n$$

This system was slightly modified by P. J. Flory, who placed the emphasis on the mechanisms of the polymerisation reactions. He reclassified polymerisations as step reactions or chain reactions corresponding approximately to condensa-tion or addition in Carother's scheme, but not completely. A notable exception occurs with the synthesis of polyurethanes, which are formed by reaction of iso-cyanates with hydroxy compounds and follow 'step' kinetics, but without the elimination of a small molecule from the respective units Reaction.

$$n\,OCN{-}R{-}NCO + n\,HO{-}R^1{-}OH \longrightarrow$$

$$OCN{-}[R{-}NHCO_2{-}R^1{-}]_n\,OH$$

In the first of these, the kinetics are such that there is a gradual build up of high relative molar mass material as reaction proceeds, with the highest molar mass molecules not appearing until the very end of the reaction. On the other hand, chain reactions, which occur only at a relatively few activated sites within the reacting medium, occur with rapid build up of a few high relative molar mass molecules while the rest of the monomer remains unreacted.

When formed, such macromolecules stay essentially unchanged while the rest of the monomer undergoes conversion. This means that large molecules appear very early in the polymerisation reaction, which is characterised by having both high relative molar mass and monomer molecules present for most of the duration of the reaction. Step reactions can give molecules having a variety of morphologies from the simple unbranched linear to the heavily crosslinked network. The final structure depends on the number of functional groups in the parent monomers - the greater the proportion with a functionality of greater than two, the more extensive will be the branching until, at sufficient degrees of branching, a highly crosslinked network emerges.

Chain reactions, by con-trast, give only linear or lightly branched polymers. Thus, in terms of the thermoplastic/thermoset classification, chain reactions give thermoplastics, while step reactions may give either thermoplastics or thermosets.

COMMON POLYMERS

A list of common polymers with their constitutional repeating units. Most polymers are named polyx. If 'x' is a single word, the name of the polymer is written out directly, as in the case of polystyrene. However, if 'x' consists of two or more words, parentheses should be used, as in the case of poly(methyl methacrylate). There are two different systems for naming polymers. The structure-based names are rigorous but seldom used in practice. They are simply given as poly(constitutional repeating unit) and the rules of nomenclature for the constitutional repeating unit are no different from those of any other organic substance. The source-based names attach 'poly' to the name of the monomer. Polystyrene is a source-based name.

Its structure-based equivalent is poly(1-phenylethylene). Polyethylene (source-based name) is denoted polymethylene according to the structure-based system. The names of polymers treated in this book are almost exclusively related to the source-based system. Most polymers have abbreviated names. It is clearly acceptable to use abbreviations in scientific papers and technical reports, but the full name of the polymer should be given the first time it appears in the text. There are numerous other names of polymers which are frequently used. 'Nylon', 'Kevlar' and 'Vectra' are a few well-known examples. The source-based or structure-based names are preferred over these truly trivial names.

Molar Mass

The enormous size of polymer molecules gives them unique properties. The influence of molar mass on the melting point of polyethylene.

Low molar mass substances (oligomers) show a strong increase in melting point with increasing molar mass, whereas a constant melting point is approached in the polymer molar mass range. Other polymers show a similar behaviour. Properties such as fracture toughness and Young's modulus show a similar molar mass dependence with almost constant values being approached in the high molar mass region. The polymer properties are often obtained in the molar mass range from 10 000 to 30 000 g mol^{-1}. Rheological properties like melt viscosity show a progressive strong increase with increasing molar mass even at high molar masses. High molar mass polymers are therefore difficult to process but they yield final products with very good mechanical properties.

There is currently no polymerisation method available that yields a polymer with molecules of a single size. Variation in molar mass among the different molecules is characteristic of all synthetic polymers. The molar mass distribution ranges over three to four orders of magnitude for many polymers. Monodisperse polymers can however be found in the nature. A full representation of the molar mass distribution is currently achieved only with size exclusion chromatography, naturally with a number of experimental limitations. Other methods yield different averages.

Polymerisation

The polymerisation process can in simple terms be divided into step-growth and chain-growth polymerisation. A typical example of step-growth polymerisation is the formation of a polyester from a hydroxy-carboxylic acid.

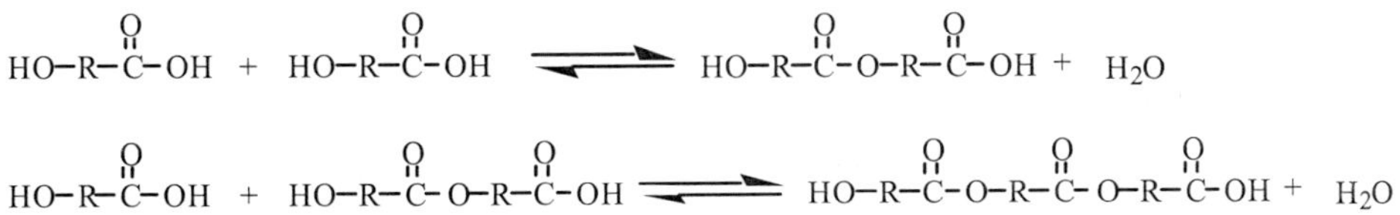

The kinetics of this polymerisation is not appreciably affected by the size of the reacting species. The number of reactive groups, in this case hydroxyl groups and acid groups, decreases with increasing length of the molecules. At any given moment, the system will consist of a mixture of growing chains and water. One problem is that all the reactions are reversible, with equilibrium being established for each reaction. The consumption of reactive groups and the formation of a high molar mass polymer require the removal of water from the system. The degree of polymerisation $\overline{X}_n$ is in the case of equal molar amounts of acid and alcohol groups equal to $1/(1-p)$, where p is the degree of conversion of reactive groups. The following Xn values are obtained for the following conversions: $p = 0.10$, $\overline{X}_n = 1.1$; $p = 0.90$, $\overline{X}_n = 10$; $p = 0.99$, $\overline{X}_n = 100$; $p = 0.999$, $\overline{X}_n = 1000$; $p = 0.9999$, $\overline{X}_n = 10\,000$. Polymers with high molar mass can thus only be obtained by running the reaction to a very high degree of conversion of the reactive groups.

Step-growth polymerisation is involved in the formation of e.g. polyesters and polyamides. Different techniques are available for obtaining a high degree of conversion and high molar mass.

If an acid chloride is used instead of the carboxylic acid, HCl is formed instead of water. The former is more easily removed from the system and higher yields and higher molar masses are obtained. The long reaction time needed to reach a high conversion is a considerable disadvantage with step-growth polymerisation.

Polymers with different molecular architectures can be made using monomers of different functionality. Tri-functional monomers yield branched and ultimately cross-linked polymers.

Chain-growth polymerisation involves several consecutive stages: initiation, propagation and termination. Each chain is individually initiated and it grows very rapidly to a high molar mass until its growth is terminated. At a given time, there are essentially only two types of molecules present: monomer and polymer. The number of growing chains is always very low. Chain-growth polymerisation is divided into several subgroups depending on the propagating site: radical, anionic, cationic or coordination polymerisation.

The initiation is often accomplished by thermal- or radiation-initiated degradation of an organic peroxide or similar unstable compound (initiator; denoted I). Free radicals are generated which attack the double bond of the unsaturated monomer (typically a vinyl monomer: CH_2=CHX) and the radical centre is moved to the end of the "chain".

It is possible that a propagating chain radical abstracts a hydrogen from an adjacent polymer molecule and that the reactive site is moved from one polymer molecule to another (chain transfer). This leads in most cases to the formation of a molecule with a long-chain branch.

The chain transfer reaction may also be intra-molecular which is the well-known 'back-biting' mechanism giving short-chain branches (primarily butyl branches) in high-pressure polymerised polyethylene. These branches, typically at a concentration of one in hundred main chain carbon atoms, reduces the crystallinity from 70-80% (linear polyethylene) to 50% in the branched polyethylene produced at high pressure. Poly(vinyl chloride) is another polymer made by radical polymerisation that have short-chain branches due to intra-molecular chain transfer.

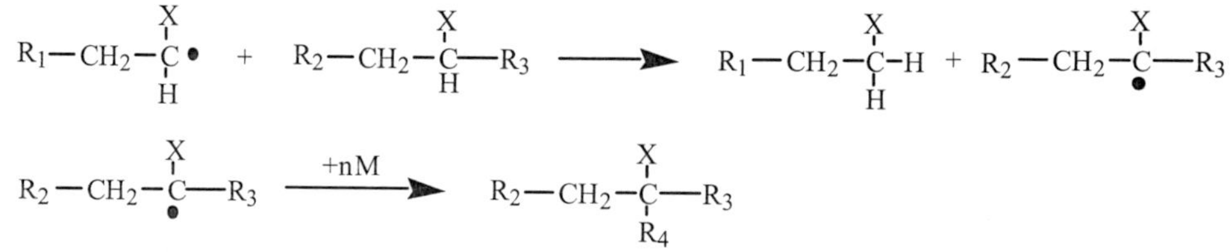

The propagation is stopped either by combination or by disproportionation:

$$R_1M\bullet + \bullet MR_2 \longrightarrow R_1MMR_2 \text{ (combination)}$$

$$R_1CH_2-\overset{X}{\underset{H}{C}}\bullet + \bullet\overset{H}{\underset{X}{C}}-CH_2R_2 \longrightarrow R_1CH_2-\overset{X}{\underset{H}{C}}H + \overset{H}{\underset{X}{C}}=CH_2R_2$$

(disproportionation)

Both anionic and cationic polymerisation includes both initiation and chain-wise propagation but with one important difference from radical polymerisation: in neither case is there any natural termination reaction. In both cases, it is possible to achieve the conditions for living polymerisation, where all chains are constantly growing until all monomer is consumed and the molar mass distribution of the resulting polymer is narrow.

It is also possible to add a new monomer and to prepare exact block-copolymers. The presence of small traces of an impurity, e.g. water, leads to chain transfer reactions and termination of the growing polymer chains. This is very different from radical polymerisation, which is very sensitive to the presence of oxygen (O_2) but essentially unaffected by traces of water.

Karl Ziegler (1955) and Guilio Natta (1955) discovered coordination polymerisation in the 1950s. This technique makes it possible to produce stereoregular polymers like isotactic polypropylene and linear polyethylene. A Ziegler-Natta catalyst requires a combination of the following substances: (i) a transition metal compound from groups IV -VIII, (ii) an organometallic compound from groups I-III, and (iii) a dry, oxygen-free, inert hydrocarbon solvent.

Commonly used systems have involved aluminium alkyls and titanium halides. The polymerisation is often rapid and exothermic, requiring external cooling. The polymer normally precipitates around the catalyst suspension. The initial work of Ziegler and Natta involved unsupported catalysts whereas many of the commercial polymers made by coordination polymerisation are today accomplished by supported catalysts. From a mechanistic point of view they are of the same class as the Ziegler-Natta catalysts. The pioneering work was due to Hogan and Banks (1958) using activated chromic oxides on silica supports. This development led to the polymerisation of linear polyethylene and linear low-density polyethylene (copolymers of ethylene and higher 1-alkenes).

The pioneering work of Kaminsky and Sinn (1980) has led to the development of a new class of catalysts often referred to as metallocenes. They consist of a bent Ti, Zr or Hf complex with two cyclopentadienyl ligands and two halide or alkyl ligands). The metallocene is often combined with a cocatalyst. The metallocenes consist of single kind of active site for polymerisation, which offer some distinct advantages over the multi-site Ziegler catalysts.

The most important feature of the metallocene catalysts is thus the control of polymer structure and properties by variation of catalyst structure, allowing the production of new polymers, unavailable by Ziegler catalysts. The metallocene technology is currently very important for the industry and new polyolefins with controlled molar mass distribution, stereostructure and comonomer distribution will enter the market.

POLY(ETHYLENE)

This polymer has one of the simplest molecular structures ($[CH_2CH_2-]_n$) and is at present the largest tonnage plastic material, having first been produced commercially in 1939 for use in electrical insulation. There is a difficulty over the nomenclature of this polymer. The IUPAC recommended name for the monomer is ethene, rather than the older ethylene.

Hence the IUPAC name for the polymer is poly (ethene). However, this name is almost never used by chemists working with the material; throughout this book, therefore, this polymer will be referred to by its more widespread name, poly(ethylene).

There are four different industrial routes to the preparation of poly (ethylene), and they yield products having slightly different properties.

The four routes are:

1. High Pressure Processes,
2. Ziegler Processes,
3. The Phillips Process,
4. The Standard Oil (Indiana) Process.

The first group of these use pressures of 1000-3000 atm and temperatures of 80-300 °C. Free-radical initiators, such as benzoyl peroxide or oxygen, are generally used and conditions need to be carefully con-trolled to prevent a runaway reaction, which would generate hydrogen, methane, and graphite rather than polymer.In general, high-pressure processes tend to yield lower density poly(ethylenes), typically in the range

0.915-0.945 g cm^{-3}, which also have relatively low molar masses. Ziegler processes are based on co-ordination reactions catalysed by metal alkyl systems. Such reactions were discovered by Karl Ziegler in Germany and developed by G. Natta at Milan in the early 1950s.A typical Ziegler-Natta catalyst is the complex prepared from titanium tetrachloride and triethylaluminium. It is fed into the reaction vessel first, after which ethylene is added. Reaction is carried out at low pressures and low temperatures, typically no more than 70 °C, with rigorous exclusion of air and moisture, which would destroy the catalyst.

The poly(ethylenes) pro-duced by such processes are of intermediate density, giving values of about 0.945 g cm^{-3}. A range of relative molar masses may be obtained for such polymers by varying the ratio of the catalyst components or by introducing a small amount of hydrogen into the reaction vessel.Lastly the Phillips and the Standard Oil (Indiana) Processes both yield high density poly(ethylenes), using relatively low pressures and temperatures.

The polymer is a very familiar material in the modern world. It is a waxy solid which is relatively low in cost, easily processed, and shows good chemical resistance. Low relative molar mass grades have the disadvantage that they suffer from so-called 'environmental stress cracking', that is they suddenly fail catastro-phically for no apparent reason after exposure to sunlight and moisture.

Despite this drawback, the various grades of poly(ethylene) have a wide range of uses. These include pipes, packaging, components for chemical plant, crates, and items for electrical insulation.

Poly(propylene)

This polymer, which has the structure $[-CH_2CH(CH_3)-]_n$ arose as a commercial material following the work of Natta on catalysts for the preparation of high relative molar mass polymers from alkenes.

Following his work on the polymerisation of ethylene, Natta showed in 1954 that it was possible to prepare analogous polymers of propylene.

Table. Details of Processes for Preparing Poly(ethylene)

Process	Catalyst	Pressure/ atm	Temperature /°C	Density of product/ g cm^{-3}
Phillips	5% CrO_3 in finely divided silica/alumina	15–35	130–160	9.6
Standard Oil (Indiana)	Supported MoO_3 with Na, Ca metal or hydride promoters	40–80	230–270	9.6

Commercial exploitation followed rapidly, and poly (propylene) was first marketed in 1957. Like poly(ethylene), there are formal problems with the nomenclature of this polymer, since its IUPAC name, poly(propene), is also rarely if ever used by polymer chemists. Since, in practice, no ambiguity is associated with the non-systematic name, this is the one that is generally used, as it will be throughout this book.When poly(propylene) was first made, it was found to exist in two possi-ble forms.

One was similar to poly(ethylene), but had greater rigidity and hardness; the other was found to be amorphous and of little strength. The first of these is now known to be isotactic, that is with a regular stereochem-istry at each alternating carbon atom. The other is now known to be atactic, that is with a random distribution of different stereochemical arrangements at each methyl-bearing carbon atom.

Commercial poly(propylene) is usually about 90-95% isotactic and is very similar to poly(ethylene) with the following exceptions:

1. It has a lower density (0.90 g cm^{-3}).
2. It has a higher softening point and hence can be used at higher temperatures. For example, it is used to make kettles, generally of the jug-type of design, and it is found to be well able to withstand the effects of exposure to boiling water.
3. It is not susceptible to environmental stress cracking.
4. It is more readily oxidised. This is a consequence of the relatively readily fragmented tertiary C—H bond within the molecule. Another inter-esting property of isotactic poly(propylene) is that it can be repeatedly flexed without causing weakness. Hence it is used for one-piece mouldings, such as boxes for card indexes, which can be used over many years without damage.

Poly(Methyl Methacrylate)

This polymer, whose name is often abbreviated to PMMA, is the most impor-tant of the commercial acrylic polymers. These polymers are formally derived from poly(acrylic acid), $[-CH_2CH(CO_2H)-]_n$. In the case of PMMA, this derivation can be thought of as coming about by replacement of the tertiary hydrogen atom by a methyl group, CH_3, and by esterification of the car-boxylic acid group with methanol, CH_3OH. This gives the structure below.

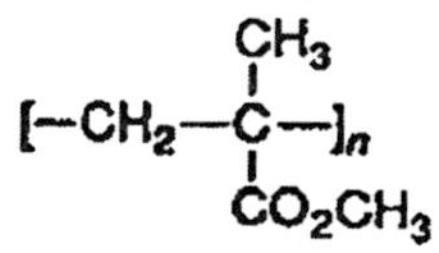

As with other major tonnage polymers, the IUPAC-recommended name for PMMA is almost never used, partly no doubt because it is extremely cumbersome. It is poly[l-(methoxycarbonyl)-l-methylethene].

Commercially PMMA is becoming relatively less important as time passes. For example, in the mid-1960s sales of PMMA ran at 40% of the total level of sales of poly(styrene); by 1980, they had fallen to 12%.

PMMA does, though, retain certain uses, including interestingly as the material from which plastic dentures are manufactured.PMMA is a transparent, glassy material, which is known from X-ray stud-ies to be amorphous at the molecular level.

It is clearly polar, as is shown by its poor insulating properties relative to poly(ethylene) and its solubility in solvents such as ethyl propanoate and trichloroethane.The main applications for PMMA arise from the combination of its transparency and its good outdoor weathering properties.

These are coupled with reasonable toughness and rigidity, so that PMMA is a useful material in a range of glazing applications. For example, it is used for display signs, street lamp fittings, and ceiling lights in factories and offices. It is also the standard material for automobile rear lamp housings.

Poly(styrene)

Poly(styrene) is a polymer which finds widespread use in the developed world on account of its desirable properties, combined with its relative cheapness. Among its features are excellent colour range, transparency, rigidity, and low water absorption.

The monomer, styrene, is a derivative of benzene, vinyl benzene. It is a colourless, mobile liquid that polymerises readily. The first report of the polymerisation reaction came in 1839, when E. Simon described the transformation of what was then called 'styrol'. He believed he had oxidised the material and called the product styrol oxide. Later, when it was realised that it contained no oxygen, the product became known as metastyrene.

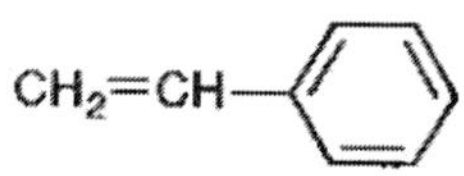

The polymer is based on a simple head-to-tail arrangement of monomer units and is amorphous, since the specific position of the benzene ring is somewhat variable and hence inhibits crystallisation. Despite its generally desirable properties, for many applications it is considered too brittle. Because of this, a number of approaches have been made to modify the mechanical properties of poly(styrene).

The most successful of these have been (i) copolymerisation and (ii) the addition of rubbery fillers.

A very large number of different copolymers of styrene have been produced and sold commercially. Among the most successful have been those polymers also containing butadiene. At the appropriate ratio of monomers, these copolymers behave as synthetic elastomers, known as styrene-butadiene rubber, SBR. Another important group of materials are the terpolymers of acrylonitrile, butadiene, and styrene, ABS. A large range of ABS materials can be made by varying not only the ratios of the three monomers but also the precise conditions of manufacture.

ABS materials generally have high impact strength and can be readily moulded to give articles of good surface finish.The alternative approach to producing strengthened styrene polymers is to use rubbery fillers. Such materials are known as high-impact poly(styrenes), a name which reflects their improved mechanical properties. They are typically prepared by dissolving the rubber in the styrene monomer, followed by polymerisation of the styrene in the usual way. The process gives a blend containing not only rubber and poly(styrene), but also small amounts of graft copolymer, comprising short poly(styrene) side chains attached to the molecules of the rubber. The overall effect of using this approach is to give materials with much better impact strengths than pure poly(styrene).Lastly, the polymerisation of styrene is used for crosslinking unsaturated polyesters. Typically a polyester is produced which contains double bonds.

Such a resin, known as a pre-polymer, is then mixed with styrene to give a viscous solution, and this solution is then worked into a glass fibre mat. The resulting composite material develops its final properties by polymerisation of the styrene, which essentially undergoes homopolymerisation, but also incorporates the occasional double bond from the pre-polymer molecule. This gives a large, three-dimensional network, intertwined with the glass reinforcement. The composite material has excellent properties and is widely used for boat hulls, shower units, and baths.

SYNTHETIC OF INFORMATIONAL POLYMERS

Pioneering work in the last decade established that nucleic acids (initially RNA) can be evolved to bind a wide variety of target ligands and catalyse many chemical transformations leading to much enthusiasm for the application of such ligands (aptamers) and enzymes (aptazymes) in therapeutic applications.

However, the natural nucleic acids, DNA and RNA, are rapidly degraded in most bodily compartments and in acidic or in alkaline conditions, respectively. Modifications to the canonical DNA/ RNA chemistry can be introduced to improve the chemical and *in vivo* stability (e.g. 2′-OMe ribonucleotides).

Such modifications have a wide range of precedents in nature and include epigenetic markers such as 5-methylcytosine or the recently discovered 5-hydroxymethylcytosine as well as other modifications, which may serve to protect viral DNA from host restriction systems such as the complete replacement of the canonical dA base with 2,6 diaminopurine in the S-2L *S. elongatus* cyanophage genome.

Other modifications include sporadic phosphorothioate modifications in the *S. lividans* genome, ribonucleotides in the rat mitochondrial genome,

5meC and pseudouridine modifications of mRNA, 2-OMe modification of rRNA as well as a myriad of tRNA modifications.

However, for therapeutic and other applications it would be desirable to evolve modified polymers *in vitro* to enable complete control over both polymer chemistry and the selection process. While chemical synthesis and replication of unnatural nucleic acids are possible, non-enzymatic polymerization has remained inefficient despite significant advances in the polymerization of some mononucleotides and oligomer units using specialised chemistries. For example, using peptide nucleic acid (PNA) pentamers, Liu and colleagues have quantitatively synthesised PNA 40-mers against a DNA backbone in one hour and performed a model selection. Despite these advances, enzymatic polymerisation using polymerases remains a desirable goal, in particular for the evolution of functional polymers.

Enzymatic polymerisation using DNA or RNA polymerases is potentially powerful but it is restricted by the stringent substrate selectivity of natural polymerases. Despite significant progress in understanding determinants of polymerase substrate specificity and the engineering of polymerases with expanded substrate spectra, most unnatural nucleotide analogues are inadequate polymerase substrates at full substitution for either synthesis and/ or as templates for reverse transcription.

A notable exception has been the use of base-modified nucleotides to generate polymers with expanded chemistry or functionalised DNA, pioneered by Famulok and colleagues.

The difficulties in engineering polymerases even for genetic polymers with similar chemistry is exemplified by the case of engineering a processive RNA polymerase from a DNA polymerase, which has proven challenging. While single incorporation catalytic rates for ribonucleotides could be improved by several orders of magnittude through the discovery of the steric gate residue in the polymerase active site, processive RNA synthesis has proved harder with reactions requiring multi-hour incubation and forcing conditions to go beyond a handful of incorporations. Indeed, the best RNA polymerase reported to date, requires an hour to extend a primer by +57 rNTP incorporations.

At the same time it is evident that highly processive RNA polymerases have evolved at least once from a DNA polymerase framework, as is the case of the single subunit RNA polymerases (ssRNAP) of both mitochondria and T-odd bacteriophages (e.g. T7 RNA pol), the polymerase domains of which derive from polA framework. We and others have therefore reasoned that there must be at least one other critical determinant in the DNA polymerase structure that aborts processive RNA synthesis in the steric gate mutants after the initial NTP incorporation. We have identified a plausible candidate for such a determinant, mutation of which enables processive synthesis of >1.5 kb mRNAs by a DNA polymerase (C. Cozens, VBP, PH; manuscript in preparation).

In that context, enzymatic synthesis of more heavily-modified genetic polymers has become a key goal of the field, particularly with modifications that increase the therapeutic potential of the synthesised polymer.

In one early success, using an engineered T7 RNA polymerase and a commercial reverse transcriptase, Burmeister and colleagues synthesised nucleic acid libraries at nearly full substitution and successfully selected aptamers against vascular endothelial growth factor (VEGF).

While a naturally occuring modification well-tolerated by natural reverse transcriptases, 2O-Me DNA is resistant to nucleases making it a suitable therapeutic aptamer.

Nucleotides in which the canonical ribofuranose sugar is replaced by an alternative ring or other structure are usually exceedingly poor polymerase substrates due to the propensity for non-canonicai helical conformations even in the direct chemical neighbourhood of ribofuranose.

A notable exception is ◎-L-threofuranosyl nucleic acids (TNA). Therminator, a A485L mutant of 9°N exo- DNA polymerase has proven useful for the generation all TNA polymers 80 – 200 nts long, which makes the use of TNA is SELEX experiments possible.

Error rates were raised compared to the use of normal dNTPs, but low enough to allow for use in SELEX. Other nucleic acid backbones, that impart greater chemical and nuclease resistance, such as HNA and LNA are even poorer substrates and even mutant polymerases like Therminator cannot incorporate more than a few nucleotides. However, through polymerase design and directed evolution, we have recently engineered polymerases with a general capacity for HNA or LNA etc. (XNA) synthesis and reverse transcription (VBP, PH, manuscript in preparation).

In conclusion, an enhanced understanding of the molecular determinants of polymerase substrate specificity together with elegant chemistry, polymerase engineering and evolution are poised to enable the synthesis and evolution of unnatural genetic polymers to yield for example aptamers composed entirely of a XNA backbone (VBP, PH, manuscript in preparation). The new sequence space opened up by these advances is likely to be a rich source of ligands and enzymes with useful applications in biotechnology, nanotechnology and material science. Beyond their practical utility these novel polymers will provide insights into the molecular parameters of the chemical encoding of information and may ultimately lead to novel organisms controlled by chemically orthogonal genetic elements or a chemically modified genome.

MODACRYLIC SYNTHETIC

A modacrylic is a synthetic copolymer. Modacrylics are soft, strong, resilient, and dimensionally stable. They can be easily dyed, show good press and shape retention, and are quick to dry. They have outstanding resistance to chemicals and solvents, are not attacked by moths or mildew, and are nonallergenic. Among their uses are in apparel linings, furlike outerwear, paint-roller covers, scatter rugs, carpets, and work clothing and as hair in wigs.

Commercial production of modacrylic fibre began in 1949 by Union Carbide Corporation in the United States. Modacrylic and acrylic fibres are similar in composition and at one time were in the same category. In 1960 the Federal Trade Commission decided to separate the two fibres and establish a category for each. The Federal Trade Commission defines modacrylic fibres as manufactured fibres in which the fibre-forming substance is any long-chain synthetic polymer composed of less than 85%, but at least 35% weight acrylonitrile units except when the polymer qualifies as rubber.

Modacrylic fibres are modified acrylic fibres made from acrylonitriles, but larger amounts of other polymers are added to make the copolymers.

The modacrylic fibres are produced by polymerizing the components, dissolving the copolymer in acetone, pumping the solution into the column of warm air (dry-spun), and stretching while hot. Modacrylics are creamy or white and are produced in tow and staple form.

If looked at in cross section views they have an irregular shape. Modacrylic fibres are also produced in many different lengths, crimp levels, deniers and they can have various shrinkage potentials. Current modacrylic fibre producers include Solutia Inc. in the U.S. and Kaneka Corporation in Japan. A modacrylic has properties that are similar to an acrylic. However, modacrylics are flame retardant and do not combust. The fibres are difficult to ignite and will self-extinguish. In addition to a modacrylic's flame retardant properties it has a relatively high durability that is comparable to wool. Modacrylic fibres have a moderate resistance to abrasion and a very low tenacity.

Modacrylics are poor conductors of heat. The fabrics are soft, warm and resilient but are prone to pilling and matting. Modacrylics display high performance when it comes to appearance retention. The fibres are quite resilient and will not wrinkle. They also have great dimensional stability and high elastic recovery, which gives them the ability to hold their shape.

Modacrylics are sensitive to loss of appearance due to improper care; therefore, it is important to know how to care for modacrylics. Modacrylics are resistant to acids, weak alkalis, and organic solvents. These fibres are also resistant to moths, mildew and sunlight.

Modacrylic fabrics can be machine washed using warm water and tumble dried on a low setting. Modacrylic pieces can also be dry-cleaned, however, they should not be steamed and should only be tumbled on cold. Some fabrics may also be cleaned using the furrier method. The fibres are heat sensitive and will shrink at 250 °F and will stiffen at temperatures over 300 °F.

Modacrylics are used primarily in applications where environmental resistance or flame retardancy is necessary or required.

Modacrylics have the ability to combine flame retardancy with a relatively low density, meaning protective gear is not uncomfortably heavy (*i.e.* shirts and trousers worn by electrical linemen). The combination of flame retardancy and low density is also useful in furnishings, draperies, and outdoor fabrics.

Modacrylics are also commonly used in fake fur fabrics, toupées, wigs and fleece-type fabric. By mixing the various forms of fibres it is easy to create a realistic fur look.

The fabrics can then be sheared or embossed to create a closer resemblance to fur. The heat-sensitivity of modacrylic also allows wigs and hairpiece to be curled and heat styled without damage.

Modacrylics are also used in fleece, knit-pile fabric backings, and nonwoven fabrics. Other uses of modacrylics include paint rollers, industrial fabrics, stuffed toys and filters.

Synthetic Fibric of Nylon

Nylon is a synthetic fabric made from petroleum products. It was developed in the 1930s as an alternative to silk, although it quickly became unavailable to civilian consumers, because nylon was used extensively during the war. Nylon, like many synthetics, was developed by Wallace Carothers at the Dupont Chemical company, which continues to manufacture it today.

Nylon is valued for its light weight, incredible tensile strength, durability, and resistance to damage. It also takes dye easily, making nylon fabrics available in a wide array of colours for consumers. Today, nylon is among the many polymer products in common daily use throughout the world. It is the second most used fibre in the United States, since, it is so versatile and relatively easy to make. Like most petroleum products, it has a very slow decay rate, which unfortunately results in the accumulation of exhausted nylon products in landfills around the world.

Nylon is made through a chemical process called *ring opening polymerization*, in which a molecule with a cyclic shape is opened and flattened. Other forms of nylon are made through the chemical reaction between two monomers: adipoyl chloride and hexamethylene diamine.

When stretched, nylon fibres even out, thin, and smooth until they reach a point at which they have no more give, yet are still very strong. Therefore, after nylon is extruded in a thread form, it is drawn or stretched after it cools to make long, even fibres. Before drawing, nylon has a tangled structure, which straightens out into parallel lines.

The strength of nylon comes from amide groups in its molecular chain, which bond together very well. Nylon also has a very regular shape, which makes it well suited to creating fabrics designed to stand up to intense forces. In fact, nylon was the primary material used in parachutes and ropes during the Second World War for this reason. It is also used for bulletproof vests and other hard wearing items.

Nylon is very sensitive to heat and should be washed and dried on cool settings. Nylon can also be hung dry, and it is favored by campers because it dries very rapidly. Nylon is a flexible textile, and as a result, it appears in a wide range of applications, from clothing to climbing equipment. Depending on how it is processed, nylon can be formed into the gossamer-like threads used in stockings or into thick toothbrush bristles.

Olefin Fibre

Olefin fibre is a synthetic fibre made from alkenes. It is used in the manufacture of various textiles as well as clothing, upholstery, wallpaper, ropes, and vehicle interiors.

Olefin is also referred to as polypropylene, polyethylene, or polyolefin. Olefin's advantages are its strength, colourfastness and comfort, stain, mildew, abrasion and sunlight resistance, and good bulk and cover.

Olefin fibres have great bulk and cover while having low specific gravity. This means "warmth without the weight." The fibres have low moisture absorption, but they can wick moisture and dry quickly. Olefin is abrasion, stain, sunlight, and chemical resistant. It does not dye well, but has the advantage of being colorfast.

Since, Olefin has a low melting point, textiles can be thermally bonded. The fibres have the lowest static of all manufactured fibres and a medium luster. One of the most important properties of olefin is its strength. It keeps its strength in wet or dry conditions and is very resilient. The fibre can be produced for strength of different properties.

Production Method

The Federal Trade Commission's official definition of olefin fibre is "A manufactured fibre in which the fibre forming substance is any long-chain synthetic polymer composed of at least 85% by weight of ethylene, propylene, or other olefin units"

Polymerization of propylene and ethylene gases, controlled with special catalysts, creates olefin fibres. Dye is added directly to the polymer before melt spinning is applied. Additives, polymer variations and different process conditions can create a range of characteristics.

High pressure production, which uses ten tons per square inch, creates a film for molded materials. Low pressure production uses a low temperature with a catalyst and hydrocarbon solvent. This process is less expensive and produces a polyethylene polymer more suitable for textile use.

The polymer is then melted, spun into water, or air cooled. The fibre is drawn out to six times the spun length. Gel spinning is a new method in which a gel form of polyethylene polymers is used.

Polyester

Polyester is a manufactured product made from synthesised polymers. It tends to be very resilient, quick drying, resistant to biological damage such as mold and mildew, easy to wash, and able to hold forms well. While polyester is often maligned as a textile, it has many useful applications. Polyester is, however, highly flammable, so care should be taken when wearing it. Most synthetic fabrics are subject to flammability by nature, because they are made from polymers.

Polyester is not just a textile. It is made from polyethylene terephthalate (PET), the same material used to make plastic drink bottles. Many drink bottles are recycled by being reheated and turned into polyester fibres, which in addition to being an efficient use, also helps keep polymers out of landfills.

Polyester is a plastic, invented in Britain in the early 1940s by two gentlemen working with synthetic polymers. In the 1950s, the new textile took off, becoming popular for its easy care, drape, and versatility.

In the United States, Dupont is one of the major manufacturers of PET and has been making Dacron polyester since, 1950. Dupont has researched the properties of polyester extensively and devised a number of uses well beyond textiles, such as mylar and other polyester films.

Dupont has also been driven to find ways to recycle polyester products to help keep costs down, as the base chemical substances that combine to form polyester are petroleum products.

To make polyester, ethylene glycol and dimethyl terephthalate are mixed together. The chemical reaction results in bisterephthalate. This substance is heated to 270° Fahrenheit (132° Celsius), where it reacts again to form polyethylene terephthalate.

Like many chemical reactions that result in polymers, the polyester making process results in unhealthy off gassing, and protection should be worn while making PET. After synthesizing the polymers, the manufacturer decides what to do with them.

PET can be formed into plastics that can later be recycled. It is a highly malleable material and appears in all sorts of applications, from drink bottles to floppy disk liners.

PET can also be used to make polyester fibre, which is used in auto upholstery, quilt batting, and clothing of all sorts.

To make polyester fibre, an extruder is used to produce very fine threads of PET. Polyester clothing tends to be slippery and silky in feel, although it can cause skin irritation for some wearers.

The polyester fibres used to make clothing can be knitted or woven, although most are knit, to maximize the flexibility of polyester. Some polyester is blended with other fabrics to provide more loft or stretch, or to minimize skin irritation.

Zylon

Zylon is a trademarked name for a range of thermoset liquid crystalline polyoxazole. This synthetic polymer material is manufactured by the Toyobo Corporation. Zylon was invented and developed by SRI International in the 1980s. Like Kevlar, Zylon is used in a number of applications that require very high strength with excellent thermal stability. Tennis racquets, table tennis blades, various medical applications, and some of the Martian rovers are some of the more well known instances. More common usage is in drive belts for snow mobiles manufactured by MBL USA Corporation. IUPAC name: poly(p-phenylene-2,6-benzobisoxazole)

Vinalon

Vinalon is a synthetic fibre, produced from polyvinyl alcohol, using anthracite and limestone as raw materials. Vinalon was first developed in 1939 by Japanese scientist, Ichiro Sakurada, Yi Sung-ki and H. Kawakami Production of this fibre was delayed for WW2.

The fibre was largely ignored in Korea until Ri defected to North Korea in 1950. Trial production began in 1954 and in 1961 the massive February 8 Vinalon Complex was built in Hamhung. Its success and widespread usage in North Korea is often pointed to in propaganda as an example of the success of the *juche* philosophy.

While Hamhung remains a major production centre for vinalon; in 1998, a vinalon factory was opened up in South Pyongan.

In early 2010, Kim Jong-il himself attended a mass rally at the February 8 Vinalon Complex in Hamhung to celebrate it's reopening after 16 years of inactivity. While Kim is often seen at politial rallies or military parades, this is the first documented time he has ever attended a industrial mass rally, said an anonymous South Korean security official.

Vinalon, also known as "Juche fibre" in the DPRK, has become the national fibre of North Korea and is used for the majority of textiles, outstripping fibre such as cotton or Nylon, which are only produced in small amounts in North Korea. Other than clothing, Vinalon is also used for shoes, ropes and quilt wadding.

Vinalon is resistant to heat and chemicals but has numerous disadvantages: it is stiff, uncomfortable, shiny, prone to shrinking and difficult to dye.

Spandex

Spandex, Lycra or elastane is a synthetic fibre known for its exceptional elasticity. It is stronger and more durable than rubber, its major non-synthetic competitor. It is a polyurethane-polyurea copolymer that was invented in 1959 by chemist Joseph Shivers at DuPont's Benger Laboratory in Waynesboro, Virginia. When first introduced, it revolutionized many areas of the clothing industry.

The name "Spandex" is an anagram of the word "expands". It is the preferred name in North America; in many European countries it is referred to as "elastane". Other brand names associated with Spandex include Lycra, Elaspan (also Invista's), Creora (Hyosung), ROICA and Dorlastan, Linel (Fillattice), and ESPA (Toyobo).

Polyethylene

Polyethylene is a type of polymer that is classified as a thermoplastic, meaning that it can be melted to a liquid and remolded as it returns to a solid state. As the name implies, polyethylene is chemically synthesised from molecules that contain long chains of ethylene, a monomer that provides the ability to double bond with other carbon-based monomers to form polymers. Polyethylene is known by other, non-official names, such as polythene in the United Kingdom. In addition, it is sometimes spelled as polyethylyne, or abbreviated to simply PE.

The first laboratory creation of polyethylene occurred in 1898 by accident at the hands of Hans von Pechmann while applying heat to another compound the German chemist previously discovered — diazomethane.

Ironically, the synthesis of polyethylene via extreme heat and pressure in an industrial setting was again made by accident, but 35 years later. A few years later, another chemist employed by the same England-based chemical company devised a method to consistently produce polyethylene under the same conditions. As a result, polyethylene became the primary source of low-density polyethylene (LDPE) production in 1939.

While polyethylene is essential to the economic health of the plastics industry, most consumers readily recognise the role it plays in everyday life.

In fact, this substance is found in many ordinary household items, such as food wrap, shampoo bottles, milk containers, toys, and the common plastic bag used to tote groceries home from the store.

However, polyethylene is also present in numerous other products that contain plastic components. For instance, it is used to manufacture artificial knee and hip replacement parts, bulletproof vests, and even glassy flooring for ice skating rinks.

Polyethylene may fall under one of several types. The distinction between them is determined by its molecular weight and branching, which is affected by its crystallization. LDPE is an example of branched polyethylene since, its carbon molecules are attached to long chains of polyethylene instead of hydrogen. Otherwise, a linear structure of carbon to hydrogen occurs, which is known as high-density polyethylene (HDPE). However, further variances in structure and molecular weight produce other forms, such as ultra high molecular weight polyethylene (UHMWPE), medium-density polyethylene (MDPE), or very low-density polyethylene (VLDPE).

While polyethylene may help to make numerous useful and durable products possible, its environmental impact is cause for concern. For one thing, it does not readily biodegrade and can reside in a landfill for hundreds of years.

However, diligent recycling may significantly reduce this problem. In addition, scientists are exploring the possibility of employing *Sphingomonas*, an aerobic bacteria shown to shorten biodegrading of some forms of polyethylene to just a few months. Environmental preservation efforts have also led to the development of bioplastics, with the aim of synthesizing polyethylene from ethanol obtained from sugarcane.

NATURALLSY OCCURRING POLYMERS

There are a number of naturally occurring polymers which find technical application, including cellulose and its derivatives, starch, and rubber. In addition, a number of important biological materials, most notably the proteins, are made up of macromolecules.

Cellulose

This is a very widely available polymer, since it is the main component of the cell walls of all plants. It is a carbohydrate of molecular formula $(C_6H_{10}O_5)_n$, where n runs to thousands. The cellulose 'monomer' is D-glucose, and the cellulose molecules are built up from this substance, effectively by condensation and removal of the elements of water.D-Glucose itself is highly soluble in water, but cellulose is not.

This is essentially a kinetic phenomenon; the hydroxy-groups in cellulose would, in principle, readily form hydrogen bonds with water molecules, and hence the cellulose macromolecule would be carried off into aqueous solution. But these hydroxy-groups interact with neighbouring cellulose molecules, making it impossible for water molecules to penetrate, or solvate, the individual molecules.

Cellulose will, however, dissolve in the somewhat strange medium aqueous ammoniacal cupric hydroxide, $Cu(NH_3)_4(OH)_2$.Cellulose is a linear polymer. Despite this, it is not thermoplastic, essentially because of its extensive intermolecular hydrogen bonding which never allows the molecules to move sufficiently for the polymer to melt. Cellulose may be solubilised by treatment with sodium hydroxide and carbon disulfide. It can be regenerated by acidification of the solution. This is the basis of the production of regenerated cellulose fibre, so-called 'viscose rayon', which is a major textile fibre.

The technique is also used for the production of continuous cellulose-derived film, so-called 'cellophane' (from 'cellulose' and 'diaphane', the latter being French for transparent).

Cellulose is also commercially modified by acetylation to produce a material suitable for X-ray and cine film. Commercially cellulose ethers are also prepared, such as methylcellulose. This material is water-soluble and gives a highly viscous solution at very low concentrations. Hence it is widely used as a thickener in latex paints and adhesives, in cosmetics and for coating pharmaceutical tablets.

Starch

Starch is a widely distributed material which occurs in roots, seeds, and fruits of plants. For commercial use, corn is the principal source, though wheat and potatoes are also used. Starch is extracted by grinding with water, filtering, centrifuging, and drying, a process which yields starch in a granular form. Starch consists of two components, amylose and amylopectin. The ratio of these varies with the source of the starch, with amylopectin usually predominating and representing some 70-85 per cent of the total mass of the starch. Amylopectin is the polymeric component of starch and consists mainly of glucose units joined at the 1,4-positions. Relative molar mass tends to be very high, *e.g.* between 7 and 70 million.

A variety of modified starches are used commercially which are produced by derivatisation to give materials such as ethanoates (acetates), phosphates, and hydroxyalkyl ethers. Modified and unmodified starches are used in approximately equal tonnages, mainly in paper-making, paper coatings, paper adhesives, textile sizes, and food thickeners.

Natural Rubber

Rubber is obtained from the juice of various tropical trees, mainly the tree Hevea brasiliensis. The juice is a latex consisting of a dispersion of polymer phase at a concentration of about 35 per cent by mass, together with traces of pro-teins, sterols, fats, and salts. The rubber is obtained either by coagulation of the latex with acid, either ethanoic or methanoic, or by evaporation in air or over a flame.

The material that results from this process is a crumbly, cheese-like substance, sometimes called raw rubber or caoutchouc. In order to develop the mechanical properties that are considered characteristic of rubber, *i.e.* so-called rubberlike elasticity, this raw rubber needs further processing, and in particular lightly crosslinking.

The polymer in natural rubber consists almost entirely of cis-poly(isoprene). The molecules are linear, with relative molar mass typically lying between 300 000 and 500 000. The macromolecular nature of rubber was established mainly by Staudinger in 1922, when he hydrogenated the material and obtained a product that retained its colloidal character, rather than yielding fragments of low relative molar mass. Vulcanisation is the term used for the process in which the rubber molecules are lightly crosslinked in order to reduce plasticity and develop elasticity.

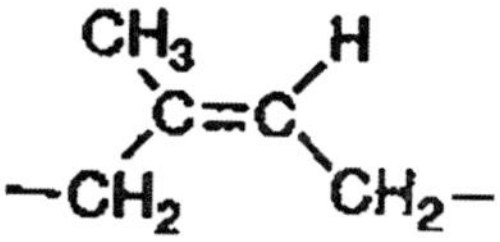

It was originally applied to the use of sulfur for this purpose, but is now used for any similar process of cross-linking. Sulfur, though, remains the substance most widely used for this purpose.Sulfur reacts very slowly with rubber, and so is compounded with rubber in the presence of accelerators and activators.

Typical accelerators are thia-zoles and a typical activator is a mixture of zinc oxide and a fatty acid.

The chemistry of the vulcanisation reactions is complicated, but generates a three-dimensional network in which rubber molecules are connected by short chains of sulfur atoms, with an average of about five atoms in each chain.A much more heavily crosslinked material can be obtained by increasing the amount of sulfur in the mixture, so that it represents about a third of the mass of the product. Heating such a mixture of raw rubber and sulfur at 150 °C until reaction is complete gives a hard, thermoset material that is not at all elastic. This material is called ebonite and is used to make car battery cases.

Proteins

The proteins are a group of macromolecular substances of great importance in biochemistry. Their very name provides testimony to this - it was coined by Mulder in 1838 from the Greek word 'proteios', meaning 'of first importance'. They appear in all cells, both animal and plant, and are involved in all cell functions. Proteins are linear polyamides formed from ?-amino acids.

An ?-amino acid is one in which carboxylic acid and the amino group reside on the same carbon atom.In nature, there are 20 amino acids available for incorporation into the protein chain. They are arranged in a specific and characteristic sequence along the molecule. This sequence is generally referred to as the 'primary structure' of the protein. Also part of the primary structure is the relative molar mass of the macromolecule.

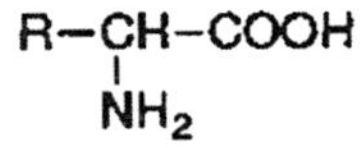

As a result of the particular amino acid sequence, the protein molecule adopts a characteristic arrangement, such as symmetrical coils or orderly foldings. This is known as the 'secondary structure'. Such individual coils or folded structures bring about a longer range stable arrangement called the 'tertiary structure'. Lastly, several protein molecules may join together to form a complex, this final grouping being known as the 'quaternary structure'.

Proteins themselves consist of large molecules with several hundred amino acids in the primary structure. Smaller units, known as polypeptides, may be formed during physiological processes. There is no clear distinction between a protein and a polypeptide; about 200 amino acid units in the primary structure is usually taken to be the minimum for a substance to be considered a protein.

The diversity in primary, secondary, tertiary, and quaternary structures of proteins means that few generalisations can be made concerning their chemical properties. Some fulfil structural roles, such as the collagens (found in bone) and keratin (found in claws and beaks), and are insoluble in all solvents.

Others, such as albumins or globulins of plasma, are very soluble in water. Still others, which form part of membranes of cells, are partly hydrophilic ('water-loving', hence water-soluble) and partly lipophilic ('lipid-loving', hence fat-soluble). Because proteins are not composed of identical repeating units, but of dif-ferent amino acids, they are nevertheless macromolecular and techniques developed for the study of true polymers have been applied to them with success.

Poly(3-hydroxybutyrate)

Poly(3-hydroxybutyrate) is a bacterial polyester that behaves as an acceptable thermoplastic, yet can be produced from renewable agricultural feedstocks and is biodegradable. It is typically produced not in the pure state,but formed alongside minor amounts of poly(3-hydroxyvalerate). The ratio of these two polymers in a given sample is determined by the ratio of glucose and propionic acid in the medium in which the bacteria live and carry out their metabolic processes.

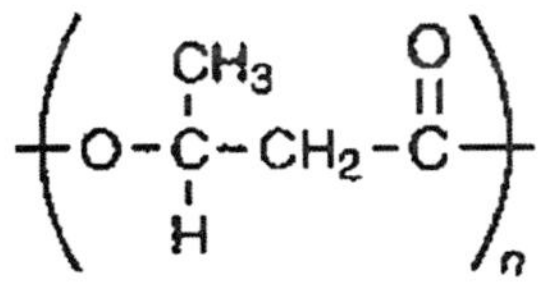

The carbon atom which carries the methyl group is chiral, but biosynthesis is stereoselective, and gives rise to a natural polymer with the R configuration. The polymer is a partially crystalline thermoplastic which melts at about 80 °C. Poly(3-hydroxybutyrate) has attracted interest as an environmentally degradable thermoplastic that can be used in packaging, agriculture and medicine. It undergoes enzymic degradation quite readily, becoming less crystalline and rapidly decreasing in molar mass. The rate of degradation depends on a number of factors, such as moisture level, nutrient supply, temperature, and pH.

Polymer Solution

The importance assigned to polymer solutions, a topic whose discussion has evolved from a mere informative mention in textbooks to whole books exclusively devoted to that subject, has become increasingly notorious. The reasons are based on key factors. In the first place, the understanding of the behaviour and both physical and chemical properties of macromolecules has been mainly sustained in studies carried out in solution, like for example the determination of the relative molecular mass, made by viscometry or gel permeation chromatography (GPC).

On the other hand, since polymer solutions are highly viscous even at low concentrations, their commercial application includes a wide range of products, from paintings to processed foods.Therefore, we can then consider polymer solutions as *liquid mixtures made of long macromolecular chains, and small, light molecules of solvent* (Grosberg and Khokhlov, 1997). This, by the way, is not a usual situation. The large size of this chains implies the employ of certain theoretical models, which should take into account, among other things, the numerous and diverse conformations that these flexible structures may assume.

This particularity is not consistent with a behaviour that could be regarded as an "ideal" behaviour. In addition to these features, it is easily understandable that the studies performed in the evaluation of the physical-chemical properties of macromolecules are focused on dilute solutions, where the chains are separated by long distances, and therefore the interaction between them is reduced to a minimum.

This is not taken for the sake of simplicity, but also because the properties of dilute solutions are governed by the properties of the individual macromolecules. In the case of concentrated solutions, the chains are entangled each other, their interaction increases, and in such conditions, the system is no longer suitable to evaluate the contribution of each macromolecule in particular.

6

Structure of Coordination Complexes

The ions or molecules surrounding the central atom are called ligands. Ligands are generally bound to the central atom by a coordinate covalent bond (donating electrons from a lone electron pair into an empty metal orbital), and are said to be coordinated to the atom. There are also organic ligands such as alkenes whose pi bonds can coordinate to empty metal orbitals. An example is ethene in the complex known as Zeise's salt, $K^+[PtCl_3(C_2H_4)]$.

Geometry

In coordination chemistry, a structure is first described by its coordination number, the number of ligands attached to the metal (more specifically, the number of σ-type bonds between ligand(s) and the central atom). Usually one can count the ligands attached, but sometimes even the counting can become ambiguous. Coordination numbers are normally between two and nine, but large numbers of ligands are not uncommon for the lanthanides and actinides. The number of bonds depends on the size, charge, and electron configuration of the metal ion and the ligands. Metal ions may have more than one coordination number.

Typically the chemistry of complexes is dominated by interactions between s and p molecular orbitals of the ligands and the d orbitals of the metal ions. The s, p, and d orbitals of the metal can accommodate 18 electrons. The maximum coordination number for a certain metal is thus related to the electronic configuration of the metal ion (more specifically, the number of empty orbitals) and to the ratio of the size of the ligands and the metal ion. Large metals and small ligands lead to high coordination numbers, e.g. $[Mo(CN)_8]^{4-}$. Small metals with large ligands lead to low coordination numbers, e.g. $Pt[P(CMe_3)]_2$. Due to their large size, lanthanides, actinides, and early transition metals tend to have high coordination numbers.

Different ligand structural arrangements result from the coordination number. Most structures follow the points-on-a-sphere pattern (or, as if the central atom were in the middle of a polyhedron where the corners of that shape are the locations of the ligands), where orbital overlap (between ligand and metal orbitals) and ligand-ligand repulsions tend to lead to certain regular geometries.

The most observed geometries are listed below, but there are many cases which deviate from a regular geometry, e.g. due to the use of ligands of different types (which results in irregular bond lengths; the coordination atoms do not follow a points-on-a-sphere pattern), due to the size of ligands, or due to electronic effects:

- Linear for two-coordination,
- Trigonal planar for three-coordination,
- Tetrahedral or square planar for four-coordination
- Trigonal bipyramidal or square pyramidal for five-coordination,
- Octahedral (orthogonal) or trigonal prismatic for six-coordination,
- Pentagonal bipyramidal for seven-coordination,
- Square antiprismatic for eight-coordination, and
- Tri-capped trigonal prismatic (Triaugmented triangular prism) for nine coordination.

Some exceptions and provisions should be noted:

- The idealized descriptions of 5-, 7-, 8-, and 9- coordination are often indistinct geometrically from alternative structures with slightly different L-M-L (ligand-metal-ligand) angles. The classic example of this is the difference between square pyramidal and trigonal bipyramidal structures.
- Due to special electronic effects such as (second-order) Jahn-Teller stabilization, certain geometries are stabilized relative to the other possibilities, e.g. for some compounds the trigonal prismatic geometry is stabilized relative to octahedral structures for six-coordination.

Isomerism

The arrangement of the ligands is fixed for a given complex, but in some cases it is mutable by a reaction that forms another stable isomer.

There exist many kinds of isomerism in coordination complexes, just as in many other compounds.

Stereoisomerism

Stereoisomerism occurs with the same bonds in different orientations relative to one another. Stereoisomerism can be further classified into:

Cis-trans Isomerism and Facial-meridional Isomerism

Cis-trans isomerism occurs in octahedral and square planar complexes (but not tetrahedral). When two ligands are mutually adjacent they are said to be cis, when opposite each other, trans. When three identical ligands occupy one face of an octahedron, the isomer is said to be facial, or fac. In a *fac* isomer, any two identical ligands are adjacent or *cis* to each other. If these three ligands and the metal ion are in one plane, the isomer is said to be meridional, or mer. A *mer* isomer can be considered as a combination of a *trans* and a *cis*, since it contains both trans and cis pairs of identical ligands.

Optical Isomerism

Optical isomerism occurs when the mirror image of a compound is not superimposable with the original compound. It is so called because such isomers are optically active, that is, they rotate the plane of polarized light. The symbol Λ (*lambda*) is used as a prefix to describe the left-handed propeller twist formed by three bidentate ligands, as shown. Similarly, the symbol □Δ (*delta*) is used as a prefix for the right-handed propeller twist.

Structural Isomerism

Structural isomerism occurs when the bonds are themselves different. Linkage isomerism is only one of several types of structural isomerism in coordination complexes (as well as other classes of chemical compounds). Linkage isomerism occurs with ambidentate ligands which can bind in more than one place. For example, NO_2 is an ambidentate ligand: it can bind to a metal at either the N atom or at an O atom.

APPLICATION OF COORDINATION COMPOUNDS

(1) They are used in photography, i.e., AgBr forms soluble complex with sodium thiosulfate in photography.

(2) $K[Ag(CN)_2]$ is used for electroplating of silver, and $K[Au(CN)_2]$is used for gold plating.

(3) Some ligands oxidise Co^{2+} to Co^{3+} ion.

(4) EDTA is used for estimation of Ca^{2+} and Mg^{2+}in hard water.

(5) Silver and gold are extracted by treating zinc with their cyanide complexes.

INCLUSION COMPOUND

In host-guest chemistry an inclusion compound is a complex in which one chemical compound (the "host") forms a cavity in which molecules of a second "guest" compound are located. The definition of inclusion compounds is very broad, extending to channels formed between molecules in a crystal lattice in which guest molecules can fit.

If the spaces in the host lattice are enclosed on all sides so that the guest species is 'trapped' as in a cage, the compound is known as a clathrate. In molecular encapsulation a guest molecule is actually trapped inside another molecule.

Cyclodextrin Inclusion Compounds

Inclusion complexes are formed between cyclodextrins and ferrocene.

When a solution of both compounds in a 2:1 ratio in water is boiled for 2 days and then allowed to rest for 10 hours at room temperature orange-yellow crystals form.

X-ray diffraction analysis of these crystals reveals a 4:5 inclusion complex with 4 molecules of ferrocene included in the cavity of 4 cyclodextrine molecules and with the fifth ferrocene molecule sandwiched between two stacks of ferrocene-cyclodextrine dimers.

Cyclodextrin also forms inclusion compounds with fragrance molecules. As a result the fragrance molecules have a reduced vapour pressure and are more stable towards exposure to light and air. When incorporated into textiles the fragrance lasts much longer due to the slow-release action.

HOW DID THE STUDY OF COORDINATION COMPOUNDS STARTED?

The coordination chemistry was pioneered by Nobel Prize winner Alfred Werner (1866-1919). He received the Nobel Prize in 1913 for his coordination theory of transition metal-amine complexes.

At the start of the 20th century, inorganic chemistry was not a prominant field until Werner studied the metal-amine complexes such as $[Co(NH_3)_6Cl_3]$.

Werner recognized the existence of several forms of cobalt-ammonia chloride. These compounds have different colour and other characteristics.

The chemical formula has three chloride ions per mole, but the number of chloride ions that precipitate with Ag^+ ions per formula is not always three. He thought only ionized chloride ions will form precipitate with silver ion.

In the following table, the number below the Ionized Cl^- is the number of ionized chloride ions per formula. To distinguish ionized chloride from the coordinated chloride, Werner formulated the Complex formula and explained structure of the cobalt complexes.

Proposed Structure of Cobalt Amonia Complexes from Number of Ionized Chloride;

Solid	*Colour*	*Ionized* Cl^-	*Complex formula*
$CoCl_36NH_3$	Yellow	3	$[Co(NH_3)_6]Cl_3$
$CoCl_35NH_3$	Purple	2	$[Co(NH_3)_5Cl]Cl_2$
$CoCl_34NH_3$	Green	1	*trans*-$[Co(NH_3)_4Cl_2]Cl$
$CoCl_34NH_3$	Violet	1	*cis*-$[Co(NH_3)_4Cl_2]Cl$

The structures of the complexes were proposed based on a coordination sphere of 6. The 6 ligands can be amonia molecules or chloride ions. Two different structures were proposed for the last two compounds, the *trans* compound has two chloride ions on opposit vertices of an octahedral, whereas the the two chloride ions are adjacent to each other in the *cis* compound. The *cis* and *trans* compounds are known as geometric isomers.

Other cobalt complexes studied by Werner are also interesting. It has been predicted that the complex $Co(NH_2CH_2CH_2NH_2)_2ClNH_3]^{2+}$ should exist in two forms, which are mirror images of each other. Werner isolated solids of the two forms, and structural studies confirmed his interpretations. The ligand $NH_2CH_2CH_2NH_2$ is ethylenediamine (en) often represented by en.

Example 1

Sketch the structures of isomers $Co(en)_3{}^{3+}$ complex ion to show that they are mirror images of each other.

Solution

The images are shown on page 242 *Inorganic Chemistry* by Swaddle. If the triangular face of the end-amino group lie on the paper, you can draw lines to represent the en bidentate ligand. These lines will show that the two images are similar to the left-hand and right-hand screws.

From the description above, sketch the structures.

Discussion

Answer these questions:

- How many isomers does the complex $Co(NH_2CH_2CH_2NH_2)_2ClNH_3]^{2+}$ have? Draw the structures of the isomers.
- How many isomers does $Co(en)_2Cl_2{}^+$ have? Sketch the structures of the isomers.
- How many isomers does $Co(NH_3)_4Cl_2{}^+$ have? Sketch the structures of the isomers.

HOW ARE COORDINATION COMPOUNDS NAMED?

Structures of coordination compounds can be very complicated, and their names long because the ligands may already have long names. Knowing the rules of nomenclature not only enable you to understand what the complex is, but also let you give appropriate names to them. Often, several groups of the ligand are involved in a complex. The number of ligand molecules per complex is indicated by a Greek prefix: mono-, di- (or bis), tri-, tetra-, penta-, hexa, hepta-, octa-, nona-, (ennea-), deca- etc. for 1, 2, 3,... 10 etc. If the names of ligands already have one of these prefixes, the names are placed in parentheses. The prefices for the number of ligands become bis-, tris-, tetrakis, pentakis- etc.

For neutral ligands, their names are not changed, except the following few:

- H_2O, aqua
- NH_3, ammine (not two m's, amine is for organic compounds)
- CO, carbonyl
- NO, nitrosyl.

Normal names that will not change;

- C_5H_5N, pyradine
- $NH_2CH_2CH_2NH_2$, ethylenediamine
- C_5H_4N-C_5H_4N, dipyridyl
- $P(C_6H_5)_3$, triphenylphosphine
- $NH_2CH_2CH_2NHCH_2CH_2NH_2$, diethylenetriamine.

The last "e" in names of negative ions are changed to "o" in names of complexes. Sometimes "ide" is changed to "o". Note the following:

- Cl^-, chloride -> chloro
- OH^-, hydroxide -> hydroxo
- O^{2-}, oxide -> oxo
- O_2^- peroxide, -> peroxo
- CN^-, cyanide -> cyano
- N_3^-, azide -> axido
- N^{3-}, nitride -> nitrido
- NH_2^-, amide -> amido
- CO_3^{2-}, carbonate -> carbonato
- $-ONO_2^-$, nitrate -> nitrato (when bonded through O)
- $-NO_3^-$, nitrate -> nitro (when bonded through N)
- S^{2-}, sulfide -> sulfido
- SCN^-, thiocyanate -> thiocyanato-*S*
- NCS^-, thiocyanate -> thiocyanato-*N*
- $-(CH_2\text{-}N(CH_2COO^-)_2)_2$, ethylenediaminetetraacetato (EDTA).

The names of complexes start with the ligands, the anionic ones first, followed with neutral ligands and the metal. If the complex is negative, the name ends with "ate". At the very end are some Roman numerals representing the oxidation state of the metal. To give and remember all rules of nomenclature are hard to do. Pay attention to the names whenever you encounter any complexes is the way to learn.

$[Co(NH_3)_5Cl]Cl_2$, Chloropentaamminecobalt(III) chloride
$[Cr(H_2O)_4Cl_2]Cl$, Dichlorotetraaquochromium(III) chloride
$K[PtCl_3NH_3]$, Potassiumtrichloroammineplatinate(II)
$PtCl_2(NH_3)_2$, Dichlorodiammineplatinum
$Co(en)_3Cl_3$, tris(ethylenediamine)cobalt(III)chloride
$Ni(PF_3)_4$, tetrakis(phosphorus(III)fluoride)nickel(0)

A bridgin ligand is indicated by placing a m- before its name. The m- should be repeated for every bridging ligand. For example:

$(H_3N)_3Co(OH)_3Co(NH_3)_3$, Triamminecobalt(III)-m-
Trihydroxotriamminecobalt(III).

Example 2

Give the structural formula for chlorotriphenyl-phosphinepalladium(II)- -dichlorochlorotri-phenylphos-phinepalladium(II).

Solution

The structure is;

```
(C6H5)3P     Cl     Cl
        \   /  \   /
         Pd     Pd
        /  \   /  \
     Cl   Cl  P(C6H5)3
```

Discussion

Answer the following questions:

- How many ionized chloride ions are there per formula?
- How many chloride ions act as bridges per formula in this complex?
- How many types of chloride ligands are there in this complex?

COMPLEX IONS AND COORDINATION COMPOUNDS

The terms *complex ion* and *coordination compound* are frequently used very loosely. Particularly is this true of the first term, which is widely employed as a catchall for many meanings. Yet most of the loose usages cannot be labeled as erroneous because of the large number of connotations that have become attached to the two terms. In a broad sense any ion in which there is more than one atom may be called a complex ion. Also, any symmetrical arrangement of atoms about a central atom, in the construction of a molecule or ion, may be designated a coordination compound. Consequently almost all inorganic substances existing either in solution or in the solid state may be described by one or both of these terms. For the purpose of simplification in the present discussion we shall define these terms in such a manner as to restrict their possible meanings, but, even so, there will occur certain exceptions to our restrictive definitions.

Definitions of Terms

As was stated in the previous paragraph, any ion which contains more than one atom, as, for example, the sulfate ion $SO_4^{=}$, may be called a complex ion. Therefore a complex ion may be defined as an electrically charged radical or group of atoms, such as $Co(NH3)_6^{3+}$, which has been formed by the direct union of simple ions or by the direct union of a simple ion with neutral molecules.

Despite the foregoing definition, stable ions such as $SO_4^{=}$, ClO_3^{-}, and NO_3^{-} are usually regarded as simple ions, inasmuch as they are incapable of dissociation into their component parts under ordinary conditions. Therefore we shall limit our definition of complex ions to those ions which are capable of some dissociation into their component parts at ordinary temperatures. In view of this interpretation, complex-ion formation may be considered to be a reaction that is more or less reversible.

Also, our concept of what constitutes a complex ion does not include lattice complexes consisting of certain double salts such as $FeSO_4(NH_4)_2SO_4 \cdot 6H_2O$ and alums such as $KAl(SO_4)^2\ 12H_2O$, since these substances are complexes only in the solid state. In water solution these compounds usually break down into simpler substances or ions.

The term complex ion carries a broader meaning than that generally assigned to coordination compound.

The latter term is usually reserved for those ions or compounds which contain coordinate linkages A coordination compound is a complex substance in which atoms or groups of atoms have been added beyond the number possible on the basis of electrovalent binding or covalent binding. In such a compound both electrons of the additional linkages are furnished by the linked atom of the coordinated group.

More simply, when a metal ion combines with an electron donor, the resulting substance is said to be a coordination compound. For example, the cupric ion combines with four ammonia molecules, in which the nitrogen atom is an electron donor, to produce the ion $Cu(NH_3)^{4++}$. This definition is not without exceptions.

Later we shall encounter coordination compounds which do not contain coordinate linkages. A typical exception is the hydrated aluminum ion, which is clearly a coordination compound, but whose components are held together by ion-dipole linkages.

The bonding between the central cation in a coordination compound and the surrounding particles may be essentially covalent or essentially ionic depending upon the nature of the components involved.

it should be pointed out that the components of a coordination compound may be held together by electrostatic forces as well as covalent linkages. Later we shall learn that the components in the ion FeF_6^{3-} are held together by ionic linkages, whereas the components of the ion $Fe(CN)_6^{3-}$ are linked with covalent bonds.

We close this section with the discouraging observation that any attempt specifically to define the terms complex ion and coordination compound or to differentiate between the two is unsatisfactory because of the many meanings which have become attached to these expressions.

Werner's Theory of Coordination Compounds

The theory of valence developed during the nineteenth century provided, for that period, a satisfactory explanation of the formation and properties of most organic compounds and the simple salts of the metals.

But this early, nonelectronic theory of valence offered no adequate explanation of the union of neutral molecules with ions to form such compounds as $CuSO_4$ $5H_2$ O and $CoCl_3$ $6NH_3$. To account for the entrance of neutral molecules into chemical reactions, Alfred Werner introduced in 1893 his concept of auxiliary valence. This concept, generally known as Werner's coordination theory was stated as follows:

Even when, to judge by the valence number, the combining power of certain atoms is exhausted, they still possess in most cases the power of participating further in the construction of complex molecules with the formation of very definite atomic linkages. The possibility of this action is to be traced back to the fact that, besides the affinity bonds designated as principal valencies, still other bonds, called auxiliary valences, may be called into action.

According to Werner the combining power of an atom is divided into two spheres of attraction: the inner, or coordinate, sphere, and the outer, or ionization, sphere.

Neutral molecules or oppositely charged ions are coordinated around a central ion in the inner sphere. The number of such groups which may be arranged around the central ion is the coordination number of the ion. He recognized that this arrangement of molecules or ions around the central ion was such as to produce a symmetrical spatial grouping. The coordination numbers which permit spatial symmetry are 2, 3, 4, 6, and 8, of which the values 4 and 6 are most common. On the basis of his auxiliary-valence theory Werner assigned to $CoCl_3^{-6}$ NH_3.

The addition of six molecules of ammonia takes place by means of auxiliary valencies.

In this complex union the chemical characteristics not only of the ammonia but also of the cobalt ion are lost. On the other hand the chloride ions retain their typical physical and chemical properties. To distinguish between the two spheres of attraction, Werner introduced the square brackets [ ] to enclose those atoms making up the coordination complex and which are therefore not ionized.

Not only may neutral molecules be coordinated around a central ion, but ions may also be coordinated in a like manner. Consequently the total electrovalent charge on the complex varies with the number of charged ions which may enter into the coordination sphere to replace neutral molecules. Coordinated molecules may be partially or completely replaced by ions, resulting in a change in the electrovalence of the complex ion. The ammines of quadrivalent platinum illustrate these variable combinations. Quadrivalent platinum has a coordination number of 6, and the ammine groups may be replaced by chloride ions to change the valence and nature of the complex in successive steps

Some Complex Ions Containing Platinum

Formula	Cation	Anion
$[Pt(NH_3)_6]Cl_4$	$[Pt(NH_3)_6]^{4+}$	$4Cl^-$
$[Pt(NH_3)5Cl]Cl_3$	$[Pt(NH_3)5Cl]^{3+}$	$3Cl^-$
$[Pt(NH_3)4Cl_2]Cl_2$	$[Pt(NH_3)4Cl_2]^{++}$	$2Cl^-$
$[Pt(NH_3)3Cl_2]Cl$	$[pt(NH_3)3Cl_3]^+$	Cl^-
$[Pt(NH_3)2Cl_4]$	Nonelectrolyte	
$[Pt(NH_3)Cl_5]K$	K^+	$[Pt(NH_3)Cl_5]^-$
$[PtCl_6]K_2$	$2K^+$	$[PtCl_6]^-$

When an ion replaces a neutral molecule in the inner sphere of a complex, the ion loses its electrovalent character and becomes attached to the central ion by a nonionized linkage. precipitating the electrovalent chloride ions with silver ions. In the first complex listed in this table there are four chloride ions which may be precipitated by silver ions. In the second complex only three of the chlorides are precipitated by silver ions, and in the third and fourth complexes only two and one electrovalent chloride ions remain. Complex 5 is a nonelectrolyte, which is almost insoluble in water and consequently, produces no ions in solution. In complexes 6 and 7 the chlorine atoms are bonded by nonionized linkages and cannot be precipitated as simple ions.

One of the outstanding features of Werner's theory was the assigning of definite positions in space to the groups in a coordination complex. Not only did it provide an explanation of isomerism that existed among known complex compounds, but it also predicted the existence of isomers of types which had not previously been observed. As an illustration of this phase of his coordination theory Werner showed that the neutral complex of divalent platinum, $[Pt(NH_3)_2Cl_2]$, exists in the form of cis-trans isomers.

The existence of these isomers also established the proof of the geometric structure of these substances. If the four groups were attached in space at the points of a regular tetrahedron, isomerism would not be possible because the four groups could be interchanged without altering their relation to one another. Consequently, from the fact that isomerism does exist, it may be concluded that the coordinating groups are arranged about the central atom of platinum within one plane

COORDINATION COMPOUNDS IN ISOMERISM

Of frequent occurrence in organic chemistry is the existence of two or more chemical compounds with identical chemical compositions but with different properties. Compounds of this description are called isomers, and the structural difference which produce dissimilar properties is referred to as *isomerism.*

The phenomenon of isomerism is not limited to the chemistry of carbon compounds but is also found less frequently among inorganic substances. The variety of chemical linkages and complexity of stereochemical relationships which occur in coordination compounds are favourable to those structural arrangements which lead to isomerism.

Moreover the multiplicity of molecular arrangements which is found among these compounds produces more types of isomerism than are possible in strictly organic molecules. In the following sections are described various types of isomerism that have been predicted by the Werner theory.

Ionization Isomerism

Ions may lose their electrovalent properties and enter into the coordination sphere. This shifting from the outer to the inner sphere is also a transition from an ionic linkage to a covalent bond, with an attendant change in properties of the atoms concerned. Hence it is possible to have two compounds with the same empirical formula, but containing different ionic groupings.

For example, the violet bromopentammine sulfate of cobalt with a formula of $[Co(NH_3)_5Br]SO_4$ is an ionization isomer of the red sulfatopentammine bromide with a formula of $[Co(NH_3)_5SO_4]Br$. Both these compounds have the same empirical formula, but the first gives a precipitate of $BaSO_4$ with $BaCl_2$ and no precipitate with $AgNO_3$ and the second gives a precipitate of AgBr with $AgNO_3$ and no precipitate with $BaCl_2$. Other illustrations of ionization isomerism are

$[Pt(NH_3)_4(OH)_2]SO_4$ and $[Pt(NH_3)_4SO_4](OH)_2$

$[Co(NH_3)_5NO_3]SO_4$ and $[Co(NH_3)_5SO_4]NO_3$.

Coordination Isomerism

In this type of isomerism two compounds of identical composition are made up of different complex ions as a result of different groups being coordinated about the same coordination centres. In the compounds

$[Co(NH_3)_6][Cr(CN)_6]$ and $[Cr(NH_3)_6][Co(CN)_6]$

cobalt occurs in the cation of the first and in the anion of the second, whereas for chromium the situation is reversed. Other examples of coordination isomerism are

$[Co(NH_3)_6][Cr(C_2O_4)_3]$ and $[Cr(NH_3)_6][CO(C_2O_4)_3]$

$[Co(en)_3][Cr(CN)_6]$ and $[Cr(en)_3][Co(CN)_6]$

Another form of the same type of isomerism may occur when the same metal appears in both cation and anion of two compounds in which there is a different distribution of donor components. Compounds exemplifying this arrangement are

$[Cr(NH_3)_6][Cr(CNS)_6]$ and $[Cr(NH_3)_4(CNS)_2]$ $[Cr(NH_3)_2(CNS)_4]$

$[Cr(NH_3)_6][Cr(C_2O_4)_3]$ and $[Cr(NH_3)_4C_2O_4]$ $[Cr(NH_3)_2(C_2O_4)_2]$

Polymerization Isomerism

This term is used to describe compounds which have the same stoichiometric composition but whose molecular compositions are multiples of the simplest stoichiometric arrangement. For example, in the three compounds

$[Pt(NH_3)_2Cl_2]$ $[Pt(NH_3)_4][PtCl_4]$ $[Pt(NH_3)_3Cl]_3]PtCl_4]$

the second and third compounds may be said to be polymers of the first. Actually, in the strict sense of the word, these are not examples of polymerization. Polymerization is usually interpreted as meaning a multiplication of simple molecules. Here we have differences in arrangements as well as a multiplication of molecular weight.

Other compounds illustrating polymerization isomerism are

$[Cr(NH_3)_3(CNS)_3]$ and $[Cr(NH_3)_5CNS]_3[Cr(CNS)_6]_2$

Hydrate Isomerism

As was pointed out the water molecule is one of the most effective coordinating agents, and as a result there are many aquo complexes. The number of molecules which may enter into the coordination sphere of a metallic ion depends upon various factors and conditions, some of which have already been discussed.

At present we are chiefly interested in the fact that the number of such molecules in the inner sphere does vary, and this variation results in the formation of hydrate isomers.

For example, there are three different hexahydrates of chromic chloride with an empirical formula of $CrCl_3 6H_2O$. One of these hydrates is violet and the other two green in colour. All three differ in the number of molecules of water in the coordination sphere. The formulas which have been assigned to these three hydrate isomers are

$[Cr(H_2O)_6]Cl_3$ $[Cr(H_2O)_5Cl] Cl_2.H2O$ $[Cr(H_2O)_4 Cl_2]Cl.2H_2O$

violet green green

These compounds differ not only in colour but also in their solubilities. Also, as might be predicted from their formulas, silver nitrate can precipitate all the chlorine from the first isomer, two-thirds from the second, and only one-third from the third.

Other compounds in which hydrate isomerism is exemplified are

$[Cr(en)_2 (H_2O)_2]Br_3$ and $[Cr(en)_2 (H_2O)Br]Br_2.H_2O$

$[Co(NH_3)_4 (H_2O)Cl]Cl_2$ and $[Co(NH_3)_4 Cl_2]Cl.H_2O$

Stereoisomerism

Stereoisomerism is one of the types of isomerism predicted by the Werner theory. Furthermore it is the most widely occurring and consequently the most important of any of the forms of isomerism. Although no attempt will be made to go into the stereochemistry of complex inorganic compounds completely, it seems worthwhile to devote more space to this phenomenon than was apportioned to the discussion of the other types of isomerism.

Stereoisomerism is a form of isomerism in which two substances of the same composition and constitution differ only in the relative positions in space assumed by certain of their constituent atoms or groups. This phenomenon is sometimes called *geometrical* isomerism, inasmuch as two isomers may possess the same geometrical figure but may differ only as to the points of attachment of coordinating groups.

Stereoisomers of compounds with the coordination numbers of 6 and 4 are well known, and in many cases proof of the configurations of these stereoisomers has been established. For the other coordination numbers no isomeric phenomena have as yet been observed to establish definite spatial structures. Hence our discussion will be centred on those stereoisomers of known existence and will omit those of speculative possibilities.

Stereoisomerism in Octahedral Complexes

The symmetrical arrangement of six coordinating groups about a central atom can be attained either in a plane configuration, as in a benzene ring, or at the corners of an octahedron. Since the number of isomers which can exist for these two spatial arrangements is different, the number of isomers observed confirm the true configuration. If the spatial arrangement is assumed to be that of a regular hexagon for a compound of the general type (Ma4 b2], where *M* indicates the central metal and a and b are different coordinating groups, then three isomers should exist, corresponding to the *ortho, meta,* and *para* derivatives of benzene.

However, the number of isomers observed for compounds of this type have always been two; consequently the possibility of a hexagonal configuration may be ruled out. Moreover, since only two isomers are possible for an octahedral arrangement, as indicated in figure such a configuration may be assumed to be the true one.

It is believed that the components of a crystal behave as closely packed spheres, and the same is probably true for the constituents of a complex ion either in a lattice or in aqueous solution. Any attempt to indicate the arrangement of groups or atoms in isomeric compounds by means of diagrams would be extremely difficult and perhaps confusing. For the sake of simplicity we shall indicate isomeric arrangements in complex ions with the simplest of geometrical figures.

By way of illustration the geometric figure for the octahedral arrangement of components in the ion with the empirical formula of $[Co(NH_{3})6]^{3+}$ could be represented by figure. However, this figure may be simplified to the form given in figure. The latter type of diagram will be used throughout our discussion to represent a complex ion with an octahedral configuration. Also, in order to obtain even greater simplicity, we shall frequently omit the symbol of the central atom.

No isomerism is possible for a complex of the type $[Ma_5b]$. However, for those complexes of the type $[Ma_4b_2]$ and $[Ma_4bc]$ two isomers are possible. These possibilities are indicated in figure and experimental findings have justified the prediction of two such isomers.

It may be observed in figure that the two coordinating groups differing from the others can occur either in the edgewise or in the axial positions.

Where one or both of the substituents occur in the edgewise position, the compound is called the *cis* isomer. If the two differing components occur in the axial position, the substance is referred to as the *trans* isomer. Cis-trans isomerism may be exemplified by the two dichlorotetrammine salts of cobalt with the formula $[Co(NH_3)_4Cl_2]Cl$.

One of these salts is green and the other violet, and they are sometimes called the *praseo* (from the Greek word for *green*) and the *violeo* salts after their colours. The praseo salt possesses the trans configuration, and the violeo salt is the cis isomer. The configurations of these two isomers are shown in figure.

Corresponding to this form of isomerism are the two dinitrotetrammine salts of cobalt with the formula $[Co(NH_3)_4(NO_2)_2]X$, where X is a halide anion. One of these salts is yellow-brown (the cis isomer), and the other one is yellow (the trans isomer).

The introduction of three univalent ions (acid radicals) into the coordination sphere of a cobaltiammine complex produces a nonionized compound. Thus, if three nitro groups are introduced into the coordination sphere of $[Co(NH_3)_6]^{3+}$, the resulting compound will be a nonelectrolyte with a formula of $[Co(NH_3)_3(NO_2)_3]$. On the basis of the octahedral theory this substance should exist in the form of two isomers. Experimentally, however, the configurations of the two isomers have not as yet been rigidly established. On the other hand it is of interest to note that only two isomers are possible for a complex with the general formula $[Ma_3b_3]$.

Again, there is one cis position and one trans position for coordinating groups of these numbers.

Optical Isomerism

The existence of mirror-image isomers is a consequence of the octahedral structure of complexes with a coordination number of 6. In fact the octahedral theory is confirmed by the occurrence of optically active isomers. *Optical isomerism* arises when two compounds exist which have configurations of atoms or groups in space about the central atom such that one structure is the mirror image of the other.

This phenomenon is also called *enantiomorphism*. An optically active substance has the power of rotating the plane of polarized light transmitted through it. The plane of the ray may be rotated to the right, in which case the substance is called the dextrorotatory (*d*) form, or it may be rotated to the left, in which case the substance is designated as the levorotatory (*l*) form. On the other hand, if the ray of polarized light is not affected, the substance is termed inactive.

Optical isomerism occurs in octahedral structures of the type $[M(AA)_3]$, in which *AA* represents a symmetrical bidentate group like ethylenediamine or the oxalate ion. Figure gives the configurations for $[Cr(C_2O_4)_3]^{3-}$, which are mirror-image forms of compounds of this type. When compounds of this kind are synthesized, they appear as inactive mixtures of the two optically active isomers and the isomers must be separated by special methods. The inactive mixtures are usually termed *racemic mixtures*.

Optical isomerism also appears in complexes with the general formulas $[M(AA)_2b_2]$ and $[M(AA)_2ab]$. These types are of even greater interest because they exist in three forms: two optically active isomers and one inactive compound. Moreover mirror-image isomerism is possible only

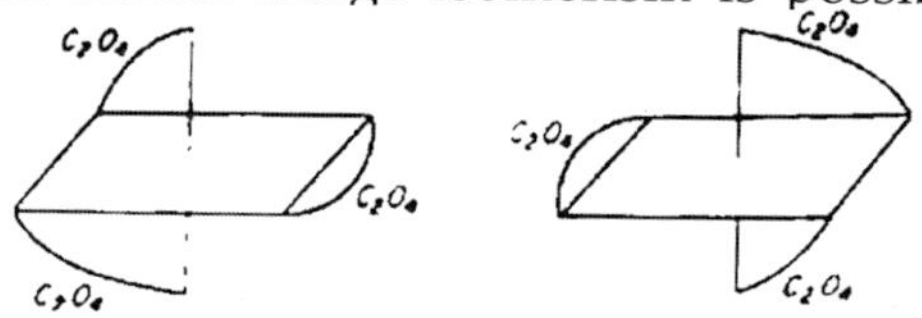

Fig. ***Mirror-image forms of*** $[Cr(C_2O_4)_3]^{3-}$.

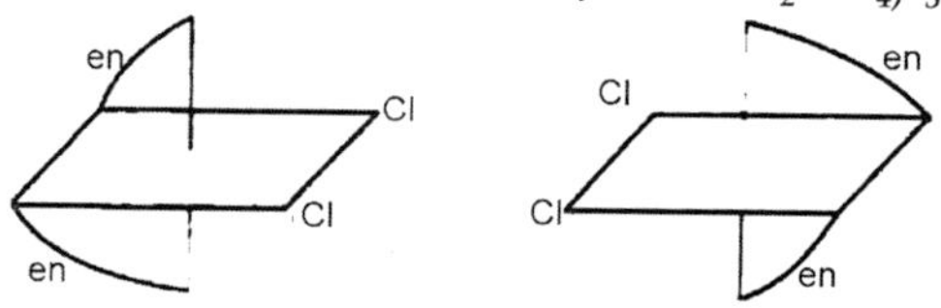

Fig. ***Optically Active Isomers of*** $[Co(en)_2Cl_2]^+$ ***with Cis Configurations.***

When the two nonchelate (unidentate) coordinating groups occur in a cis configuration. A compound rotates the plane of polarized light when it is possible to construct two molecular models which are mirror images of each other. Such models cannot be superimposed upon each other and are said to be *asymmetric*. The optically active isomers with cis configurations for $[CO(en)_2Cl_2]Cl$ are shown in figure. On the other hand, when the unidentate groups are in the trans position, the complex has a plane of symmetry and is optically inactive. The configuration for $[Co(en)_2Cl_2]^+$ with the unidentate groups in the trans position is shown in figure. The "en" in figures. is an abbreviation for ethylenediamine, $NH_2CH_2CH_2NH_2$. This organic molecule forms a bidentate grouping because of its ability to attach itself at each end to the central atom.

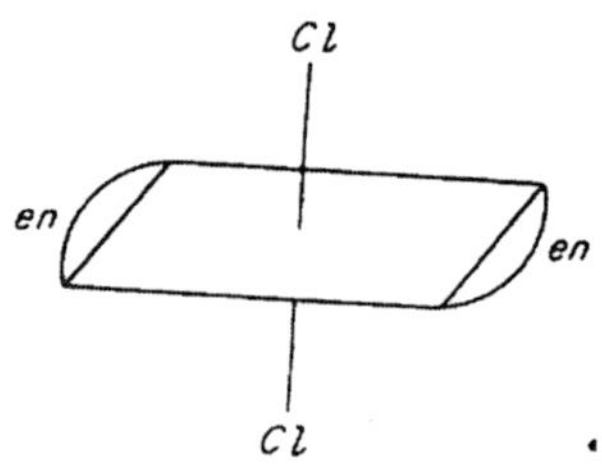

Fig. ***Inactive Isomer of*** $[Co(en)_2Cl_2]^+$ ***with Trans Configuration***

The majority of chelate compounds are produced by the multiple attachments of organic molecules to central metal atoms.

Furthermore optical isomerism and the asymmetric carbon atom are among the cornerstones of structural organic chemistry. Consequently, at first it was thought that optical activity in coordination compounds might be due to asymmetric carbon atoms within the organic molecules attached to the central atoms.

Although Werner's theory predicted optical activity among octahedral complexes with bidentate groupings, it seemed desirable to confirm the octahedral theory by preparing optically active compounds of purely inorganic nature. Two such compounds have been prepared and resolved, one by Werner [1] in 1914, and the other by F. G. Mann [2] in 1933. The resolution effected by Werner was that of the hexol dodecammine tetracobaltic ion, which has the formula of

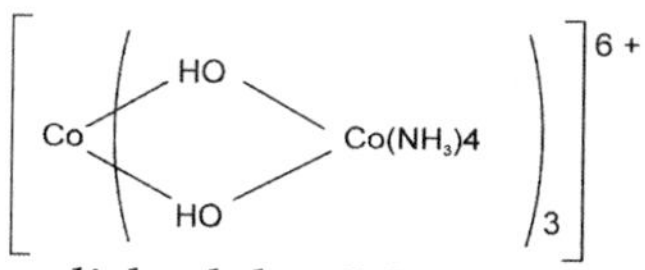

The resolution accomplished by Mann was with the sodium salt of diaquodisulfamidorhodium, which has the formula

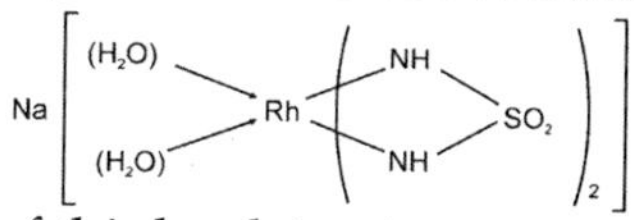

It is beyond the scope of this book to give an account of these investigations, but the preparation and resolution of optically active inorganic compounds gave brilliant confirmation to the Werner theory.

Stereoisomerism in Square-Planar Complexes

Four coordinating groups may be placed about a central metal atom in two spatial arrangements. The central atom is placed either in the centre of a tetrahedron with the coordinating groups on each corner or in the centre of a square-planar structure with the four coordination positions occupying the corners of the quadrilateral.

The two possibilities are indicated in figure. Although both spatial arrangements are found among the coordination compounds of the metals, isomerism occurs more frequently among compounds with a square-planar arrangement than among those with a tetrahedral structure.

The reasons for this disparity of isomeric forms between the two possible structures will become apparent as our discussion continues.

No stereoisomers are possible with either the tetrahedral or the planar structures for those complexes of the types $[Ma_4]$, $[Ma_3b]$, and $[Mab_3]$, because all possible arrangements of components in each of these substances are exactly the same.

In the case of materials of the type $[Ma_2b_2]$, cis and trans isomers may be predicted for the planar structure, but none are possible for the tetrahedral. The cis and trans isomeric arrangements for the planar structure are given in figure. In these arrangements the cis isomer is designated as the one with similar groups in adjacent positions and the trans isomer as the one where these groups are opposite.

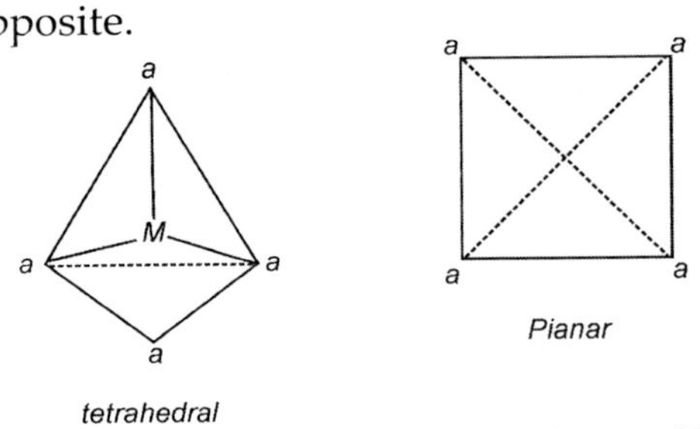

Fig. ***Spatial Arrangements for Compounds of Coordination Number 4.***

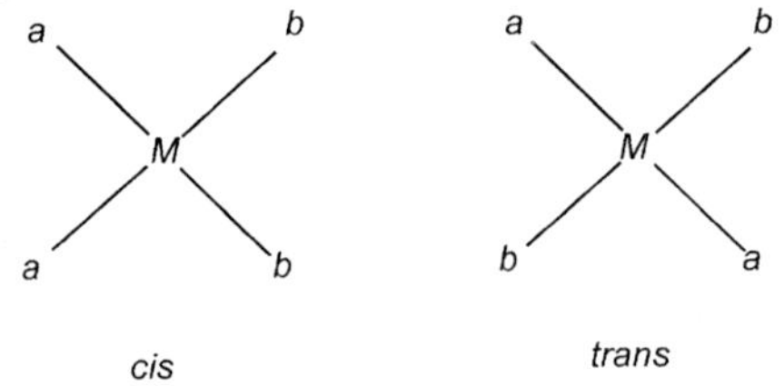

Fig. ***The Cis and Trans Isomeric Arrangements for the Planar Structure of a Compound of Type*** **[Ma $_2$ b $_2$].**

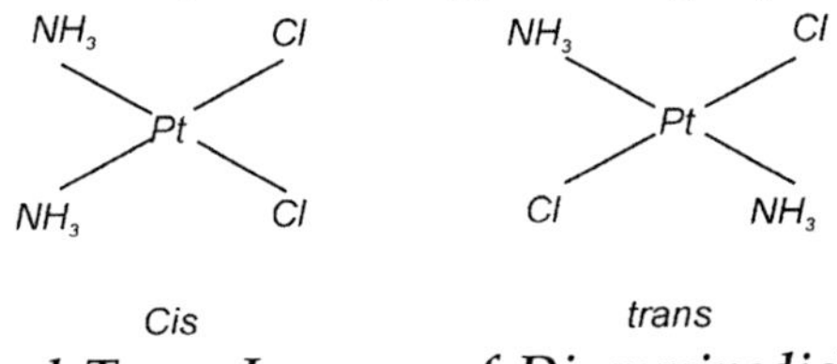

Fig. ***The Cis and Trans Isomers of Diamminedichloroplatinum.***

The existence of diamminedichloroplatinum in two isomeric forms led Werner in 1893 to propose a planar arrangement for the four valencies of the platinum atom. The spatial arrangements in these isomers are given in figure. Since that time dozens of such isomeric pairs have been studied, and proof of the planar structure for many tetracovalent metals has been obtained by X-ray studies. Particularly is this true for compounds of nickel, palladium, and platinum. By means of X-ray analyses it has been proved that the anions $[PtCl_4]^=$ and $[PtCl_4]^=$ are planar and that the four nitrogen atoms in the cation $[Pt(NH_3)_4]^{++}$ are in the same plane as the platinum atom. Additional confirmation of the planar arrangement of the valency bonds of the platinous ion came from the discovery of three isomers for the cation [Pt(NH $_3$)(NH $_2$

OH)(C_6H_5N) (NO_2)]$^+$. In addition to X-ray analysis other physical methods have been used to prove the correctness of the planar distribution. Janes 1 has measured the magnetic susceptibilities of the more important four-coordinated palladium compounds, and his results support certain planar arrangements. One of the most convincing methods of determining the structures of complex compounds is the measurement of dipole moments. Jensen 2 has demonstrated that, for a large number of platinum complexes of the type Ma 2 b 2, the dipole moments are large for cis isomers and zero for trans forms. The only possible explanation for this behaviour is a planar configuration for bivalent platinum. Probably the most unique proof of a planar distribution of the valencies about tetracovalent platinum has been obtained by Mills and Quibell. 3 Without going into a complete explanation of their ingenious method, it may be outlined briefly by saying that they succeeded in preparing an optically inactive mixture of stil-benedia-minoisobuty-lenediamino platinous salts and then resolved the tetraammine mixture into two stable optically active isomers. The general formula for the complex is

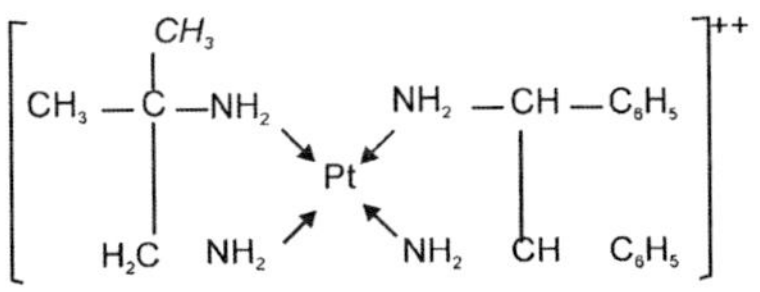

The construction of three-dimensional models of this structure reveals the fact that, if the platinum atom is tetrahedral, the complex possesses a plane of symmetry and therefore cannot be resolved into isomers. The fact that it is resolvable into stable enantioinorphs is elegant proof of the planar arrangement of the platinum valencies.

Stereoisomerism in Tetrahedral Complexes

The stereochemistry of carbon compounds indicates that in the tetrahedral arrangement it is possible to have isomerism only when all four coordinating groups are different. Thus in a compound of the type [Mabcd], when the arrangement is tetrahedral, the attachment of four different atoms or groups should result in the production of a pair of enantiomorphs (mirror-image isomers). On the other hand, with this configuration, no other type of isomerism is possible. The mirror-image isomers of this type are indicated graphically in figure.Although the tetrahedral configuration in molecules and ions such as

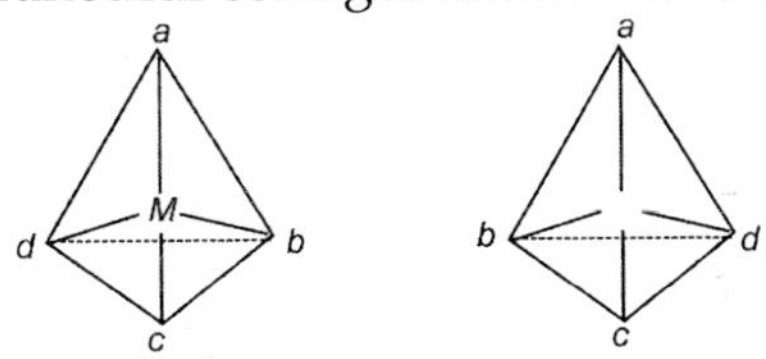

Fig. Mirror-Image Isomers of a Compound of Type [Mabcd]

(NH 3) 2 ZnCl 2, CoCl 4 =, and [Pt(CH 3) 3 Cl] is quite common, stereoisomerism is of fairly rare occurrence. It might be well to repeat that where stereoisomerism does exist in tetrahedral complexes it must necessarily be in the form of optical isomerism.

Furthermore the synthesis and resolution of asymmetric molecules or ions have almost invariably involved the presence, in the complex, of substituted hydrocarbon groups or radicals. For this reason such compounds are of interest and importance in the field of organic as well as that of inorganic chemistry.

7

Concepts in Supramolecular Chemistry

MOLECULAR SELF-ASSEMBLY

Molecular self-assembly is the construction of systems without guidance or management from an outside source (other than to provide a suitable environment). The molecules are directed to assemble through noncovalent interactions. Self-assembly may be subdivided into intermolecular self-assembly (to form a supramolecular assembly), and intramolecular self-assembly (or folding as demonstrated by foldamers and polypeptides). Molecular self-assembly also allows the construction of larger structures such as micelles, membranes, vesicles, liquid crystals, and is important to crystal engineering.

Molecular Recognition and Complexation

Molecular recognition is the specific binding of a guest molecule to a complementary host molecule to form a host-guest complex. Often, the definition of which species is the "host" and which is the "guest" is arbitrary. The molecules are able to identify each other using noncovalent interactions. Key applications of this field are the construction of molecular sensors and catalysis.

Template-directed Synthesis

Molecular recognition and self-assembly may be used with reactive species in order to pre-organize a system for a chemical reaction (to form one or more covalent bonds). It may be considered a special case of supramolecular catalysis. Noncovalent bonds between the reactants and a "template" hold the reactive sites of the reactants close together, facilitating the desired chemistry. This technique is particularly useful for situations where the desired reaction conformation is thermodynamically or kinetically unlikely, such as in the preparation of large macrocycles. This pre-organization also serves purposes such as minimizing side reactions, lowering the activation energy of the reaction, and producing desired stereochemistry. After the reaction has taken place, the template may remain in place, be forcibly removed, or may be "automatically" decomplexed on account of the different recognition properties of the reaction product. The template may be as simple as a single metal ion or may be extremely complex.

Mechanically-interlocked Molecular Architectures

Mechanically-interlocked molecular architectures consist of molecules that are linked only as a consequence of their topology. Some noncovalent interactions may exist between the different components (often those that were utilized in the construction of the system), but covalent bonds do not.

Supramolecular chemistry, and template-directed synthesis in particular, is key to the efficient synthesis of the compounds. Examples of mechanically-interlocked molecular architectures include catenanes, rotaxanes, molecular knots, and molecular Borromean rings.

Dynamic Covalent Chemistry

In dynamic covalent chemistry covalent bonds are broken and formed in a reversible reaction under thermodynamic control. While covalent bonds are key to the process, the system is directed by noncovalent forces to form the lowest energy structures.

Biomimetics

Many synthetic supramolecular systems are designed to copy functions of biological systems. These biomimetic architectures can be used to learn about both the biological model and the synthetic implementation. Examples include photoelectrochemical systems, catalytic systems, protein design and self-replication.

Imprinting

Molecular imprinting describes a process by which a host is constructed from small molecules using a suitable molecular species as a template. After construction, the template is removed leaving only the host. The template for host construction may be subtly different from the guest that the finished host bind. In its simplest form, imprinting utilizes only steric interactions, but more complex systems also incorporate hydrogen bonding and other interactions to improve binding strength and specificity.

Molecular Machinery

Molecular machines are molecules or molecular assemblies that can perform functions such as linear or rotational movement, switching, and entrapment. These devices exist at the boundary between supramolecular chemistry and nanotechnology, and prototypes have been demonstrated using supramolecular concepts.

BUILDING BLOCKS OF SUPRAMOLECULAR CHEMISTRY

Supramolecular systems are rarely designed from first principles. Rather, chemists have a range of well-studied structural and functional building blocks that they are able to use to build up larger functional architectures. Many of these exist as whole families of similar units, from which the analog with the exact desired properties can be chosen.

Synthetic Recognition Motifs

- The pi-pi charge-transfer interactions of bipyridinium with dioxyarenes or diaminoarenes have been used extensively for the construction of mechanically interlocked systems and in crystal engineering.
- The use of crown ether binding with metal or ammonium cations is ubiquitous in supramolecular chemistry.
- The formation of carboxylic acid dimers and other simple hydrogen bonding interactions.
- The complexation of bipyridines or tripyridines with ruthenium, silver or other metal ions is of great utility in the construction of complex architectures of many individual molecules.
- The complexation of porphyrins or phthalocyanines around metal ions gives access to catalytic, photochemical and electrochemical properties as well as complexation. These units are used a great deal by nature.

Macrocycles

Macrocycles are very useful in supramolecular chemistry, as they provide whole cavities that can completely surround guest molecules and may be chemically modified to fine-tune their properties.

- Cyclodextrins, calixarenes, cucurbiturils and crown ethers are readily synthesized in large quantities, and are therefore convenient for use in supramolecular systems.
- More complex cyclophanes, and cryptands can be synthesized to provide more taliored recognition properties.
- Supramolecular metallacycles are large aggregates formed from angular and linear modules. Common metallacycle shapes in these types of

applications include triangles, squares, and pentagons, each bearing functional groups that connect the pieces via "self-assembly."

Structural Units

Many supramolecular systems require their components to have suitable spacing and conformations relative to each other, and therefore easily-employed structural units are required.

- Commonly used spacers and connecting groups include polyether chains, biphenyls and triphenyls, and simple alkyl chains. The chemistry for creating and connecting these units is very well understood.
- nanoparticles, nanorods, fullerenes and dendrimers offer nanometer-sized structure and encapsulation units.
- Surfaces can be used as scaffolds for the construction of complex systems and also for interfacing electrochemical systems with electrodes. Regular surfaces can be used for the construction of self-assembled monolayers and multilayers.

Photo-/electro-chemically active units

- Porphyrins, and phthalocyanines have highly tunable photochemical and electrochemical activity as well as the potential for forming complexes.
- Photochromic and photoisomerizable groups have the ability to change their shapes and properties (including binding properties) upon exposure to light.
- TTF and quinones have more than one stable oxidation state, and therefore can be switched with redox chemistry or electrochemistry. Other units such as benzidine derivatives, viologens groups and fullerenes, have also been utilized in supramolecular electrochemical devices.

Biologically-derived Units

- The extremely strong complexation between avidin and biotin is instrumental in blood clotting, and has been used as the recognition motif to construct synthetic systems.
- The binding of enzymes with their cofactors has been used as a route to produce modified enzymes, electrically contacted enzymes, and even photoswitchable enzymes.
- DNA has been used both as a structural and as a functional unit in synthetic supramolecular systems.

PLATINUM IS USED IN TREATING CANCER

Platinum is used in jewellery, and in catalytic converters. Catalytic converters are used in car engines to convert polluting (TOXIC) gases (e.g. carbon monoxide, nitrogen oxides) into gases normally present in air (carbon dioxide, oxygen, nitrogen). Platinum acts as a CATALYST, which means that it speeds all these reactions up.

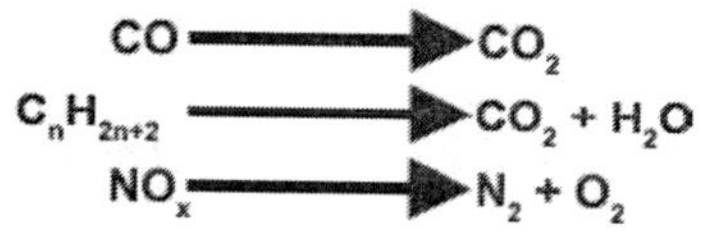

Platinium is used in an anti-cancer drug called cis-platin. Cis-platin has 2 chlorines (Cl) and 2 ammonia (NH_3) groups arranged round a platinum (Pt) atom. It is used for chemotherapy to treat cancers of the neck, ovary and testes.

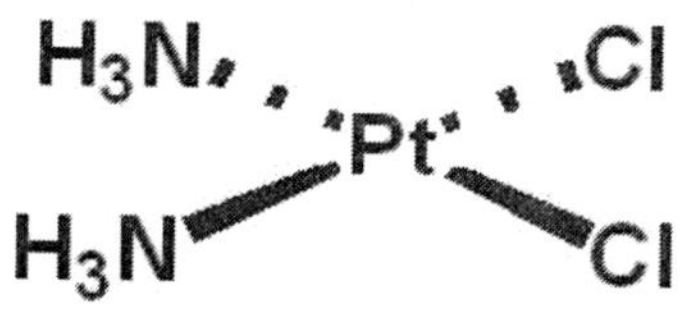

Cis-Platin

Cis and Trans?

Cis-platin has a good success rate for treating cancer. However, the other form of the molecule, *trans*-platin is inactive.

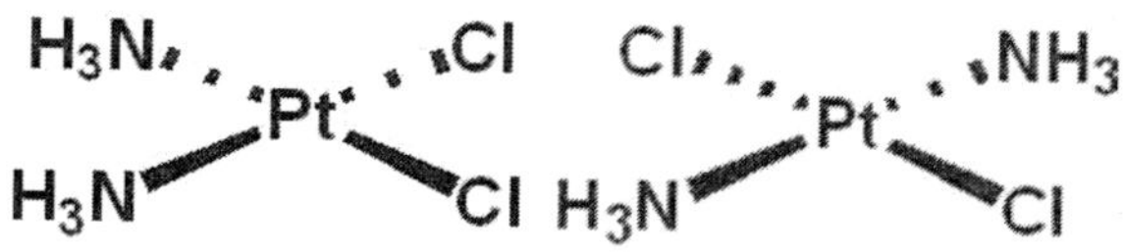

Cis- (left) and trans-platin (right)

Scientists would like to find an alternative to cis-platin as a cancer treatment as it is very toxic to the whole body and tends to get left behind in liver and kidneys. Chemists work to try and improve medicines and reduce their side effects. Researchers at the University of Edinburgh have built new molecule like cis-platin but with 6 side groups instead of 4, and it works in such a way that the toxic side effects of the chemotherapy are reduced.

The medicine is taken in an inactive form, which is then activated by shining a red laser light on the area where the cancer is. Red light penetrates skin very well and also transforms the medicine into an active form, which will treat the cancer. Therefore, the medicine is only toxic in the region of the tumour and there is less of the active form of the drug in the body, therefore reducing the side effects.

The inactive form of the molecule is on the left. The red light causes a reaction to happen producing the anti-cancer compound on the right.

TITANIUM IS USED TO MAKE HIP REPLACEMENTS

Titanium is a light metal but very strong, e.g. it is as strong as steel but 45% lighter. For these properties it is used to make lightweight alloys for aircraft, bikes and laptop computers.

The compound Titanium dioxide (TiO_2) absorbs UV light and is used in sunscreen. It is the largest use of the element because it is used extensively in paints.

Titanium is resistant to corrosion by sea water and therefore is used on parts of ships. The non-reactive properties of titanium make it ideal for use in the body as hip replacements and pacemakers, which won't be attacked by body fluids. The strength of titanium also means that it can be used in hip joints, which get a lot of wear and tear. Unfortunately titanium is expensive but we can use ELECTROPLATING (link to Material World topic) to coat cheaper metals with a thin layer of titanium.

BARIUM IS USED TO SEE INSIDE OUR BODIES

'Barium meals' and 'barium enemas' (barium sulfate) are given to patients suffering from digestive disorders. Barium absorbs X-rays and therefore when it is swallowed, it can be used to produce X-ray images of the intestines and any blockages are revealed.

But barium compounds are generally poisonous so how can it be swallowed? Barium sulfate is the compound used because it is INSOLUBLE in water and therefore cannot be absorbed by the body.

LEAD: A TOXIC METAL THAT PROTECTS US FROM X-RAYS

Some metals are toxic and we want to prevent them getting into our bodies as much as possible. One of these metals is lead. Lead can damage various systems in the body and is especially harmful to the development of young brains. But we didn't always know that lead is bad for us and it has been used for many things in the past which we are now phasing out e.g. lead in petrol (added as an 'anti-knocking agent), lead in paints (lead compounds give very vibrant colours) and lead pipes to transport water into our houses.

In the time of Elizabeth I, it was fashionable for ladies to look very pale and they used white lead compounds as make-up. This often gave them sores on their faces but off course they just put on more lead make-up to cover them up!! A few women even died of lead poisoning but this didn't stop people following the fashion!

Not all forms of lead are toxic.... lead ore which is found in the earth is not toxic and we use lead crystal glasses which are safe. Also lead glass as a shield in medical x-ray apparatus.

APPLICATIONS OF MATERIALS TECHNOLOGY

Supramolecular chemistry and molecular self-assembly processes in particular have been applied to the development of new materials. Large structures can be readily accessed using bottom-up synthesis as they are composed of small molecules requiring fewer steps to synthesize. Thus most of the bottom-up approaches to nanotechnology are based on supramolecular chemistry.

Catalysis

A major application of supramolecular chemistry is the design and understanding of catalysts and catalysis. Noncovalent interactions are extremely important in catalysis, binding reactants into conformations suitable for reaction and lowering the transition state energy of reaction. Template-directed synthesis is a special case of supramolecular catalysis. Encapsulation systems such as micelles and dendrimers are also used in catalysis to create microenvironments suitable for reactions (or steps in reactions) to progress that is not possible to use on a macroscopic scale.

Medicine

Supramolecular chemistry is also important to the development of new pharmaceutical therapies by understanding the interactions at a drug binding site.

The area of drug delivery has also made critical advances as a result of supramolecular chemistry providing encapsulation and targeted release mechanisms. In addition, supramolecular systems have been designed to disrupt protein-protein interactions that are important to cellular function.

Data Storage and Processing

Supramolecular chemistry has been used to demonstrate computation functions on a molecular scale. In many cases, photonic or chemical signals have been used in these components, but electrical interfacing of these units has also been shown by supramolecular signal transduction devices. Data storage has been accomplished by the use of molecular switches with photochromic and photoisomerizable units, by electrochromic and redox-switchable units, and even by molecular motion. Synthetic molecular logic gates have been demonstrated on a conceptual level. Even full-scale computations have been achieved by semi-synthetic DNA computers.

Green Chemistry

Research in supramolecular chemistry also has application in green chemistry where reactions have been developed which proceed in the solid state directed by non-covalent bonding. Such procedures are highly desirable since they reduce the need for solvents during the production of chemicals.

Other Devices and Gunctions

Supramolecular chemistry is often pursued to develop new functions that cannot appear from a single molecule. These functions also include magnetic properties, light responsiveness, self-healing polymers, molecular sensors, etc. Supramolecular research has been applied to develop high-tech sensors, processes to treat radioactive waste, and contrast agents for CAT scans.

METALS IN MEDICINAL INORGANIC CHEMISTRY

As a discipline, medicinal inorganic chemistry has only existed for about the last 35–40 years, since the serendipitous discovery of the antitumor activity of cisplatin, cis-[Pt(NH3)2Cl2]. Orvig and Abrams, in their introduction to an issue of Chemical Reviews devoted to Medicinal Inorganic Chemistry, cited here as reference 1, define the discipline as specifying a known chemical compound to have a specific activity. Furthermore, to meet the criteria of the discipline, studies should have been carried out to elucidate the compound's mechanism of medicinal action and to optimize and improve the compound's physiological activity. The rapid growth of understanding in bioinorganic chemistry has seen a parallel growth of interest in the medicinal properties of various metals whether or not they are found in biological species.

Inorganic or metal-containing medicinal compounds may contain either (a) chemical elements essential to life forms—iron salts used in the treatment of anemia—or (b) nonessential/toxic elements that carry out specific medicinal purposes—platinum-containing compounds as antitumor agents or technetium and gadolinium complexes as medical diagnostic tools.

Any compound or element is subject to the limitations of the so-called Bertrand diagram, which indicates the relationship among deficiency, optimum physiological response, and toxicity for the species in question. Every medicinal compound or element will have a different response curve defining its limits of deficiency, optimal response, and toxicity.

Although many elements and compounds are required at some dosage for an organism's survival, all elements and compounds may be deleterious if taken in overly large doses. Some elements and compounds are required in a certain range of concentration, while some appear to be toxic at minimal dosage levels.

Two examples of the latter are elemental lead and mercury as well as their compounds. While these elements have always been present in our atmosphere, they are especially dangerous because they form long-lasting compounds with organic species and accumulate low in the food chain. Methyl mercury compounds accumulating in fish, for instance, caused a terrible wasting disease in Japan before the toxic effects of CH_3Hgþ contamination of food was recognized. Tetraethyl lead, $Pb(C_2H_5)_4$, was used as a octane-increasing gasoline additive until dangerous accumulations of lead in the atmosphere were traced to its usage. Air pollution by lead and lead compounds remains a great problem in countries where leaded gasoline is still used extensively. Toxic elements and compounds may not be equally dangerous for all organisms at all levels.

The radioactive elements have great value in that they can be used to screen for disease and to kill tumours at given body locations. On the other hand, radioactivity is known to create mutations that may cause diseases such as leukemia and other cancers.

Lead, mercury, and thallium appear to be dangerous no matter how they appear in biological species. While chromium is an essential element required for normal carbohydrate and lipid metabolism and its deficiency can lead to adult onset diabetes, Cr(VI) uptake through anion channel transport into cells via CrO_2^{-4} causes several toxic effects. Following reduction inside cells, Cr(III) forms adducts involving the phosphate backbone of DNA, the N7 atom of guanine, amino acids such as cysteine, glutathione, and larger peptides and protein molecules.

In addition to essentiality, chemical elements and compounds found in biological systems obey homeostatic criteria. Homeostasis, the maintenance of chemical elements and compounds at optimal physiological levels, is a complex interactive system involving many elements and the biological energy supply. Interactions between deficiencies and excesses of elements and compounds interfere with homeostasis and may cause disease states in organisms. How the complex processes of homeostasis are managed within systems has been the subject of much research and writing.

The book The Biological Chemistry of the Elements: The Inorganic Chemistry of Life, written by J. J. R. Frausto da Silva and R. J. P.Williams, provides a comprehensive survey of the field. Alteration of homeostatic concentrations of metals within the human body or removal of oversupply of toxic metals has been accomplished medically by the addition of metal chelators such as ethylenediaminetetracetate (EDTA). Metal chelators, while useful, have many disadvantages, one of which is the undesired complexation and removal of other essential metals. Researchers have therefore tried to understand the processes by which essential metal cofactors such as copper or zinc ions, toxic if found in inappropriate locations or in excess concentration, are acquired by intracellular proteins such as superoxide dismutase. With this understanding, it is believed that one might specifically target and remove only the offending metal ions.

Many metals are used in medicines to cure diseases such as cancer (Platinum) and Alzheimer's disease (Gold).

METALS ARE ESSENTIAL FOR LIFE

This is a table of all the known elements arranged in order of atomic number (how many protons in the atoms of the element), in such a way that elements are arranged vertically in "groups" that have similar properties. e.g. Lithium (Li) is in the group of alkali metals and these metals all react with water.

Iron Compounds are Essential to Life

In our bodies, iron is essential for carrying oxygen in our blood to provide energy for our cells.

People who are deficient in iron are said to be anaemic. They are often very tired and lethargic and low in energy. There are many forms of anaemia, but the most common is iron deficient anaemia.

You can work out how much iron you have in your body, because everyone has approximately 0.06 g of iron per kg body weight. For example, if you weigh 60 kg, you should have about 3.6 g (0.06 x 60) of iron in your body.

Breakfast cereals are said to contain a lot of iron, but by doing a simple experiment we can show that the iron is added as iron filings, very little of which will be absorbed by the body, so don't rely on breakfast cereals for your daily supply! Take a handful of Special K and grind it into smaller pieces (by hand). Put the cereal into a bowl and add about 100 ml water. You will need a magnetic stirrer and a follower (the magnet). Stir the cereal for about 10 minutes. Lift the follower out of the mixture and you should be able to see iron filings on it.

Iron Makes Our Blood Red

In our red blood cells, we have a protein called haemoglobin. Each haemoglobin protein contains 4 iron atoms that bind to oxygen. This is how oxygen is carried round our blood stream.

Iron is what makes our blood red. Some living things have different metals in their transport system, for example, crabs & crustaceans use copper in their transport system which gives them blue blood, and sea shrimps have green blood as they use the metal vanadium.

Different metals give different colours.

How Do We Get Energy From Oxygen?

In our bodies, oxygen undergoes a reaction called a REDUCTION reaction, which gives out energy.

In the TRAFFIC LIGHT experiment, oxygen is REDUCED and a dye called indigo carmine is OXIDISED when it reacts with oxygen. When indigo carmine is oxidised, it changes colour.

As you shake a solution, air dissolves in it. As the flask is shaken, oxygen in the air dissolves in the solution and oxidises the indigo carmine from amber to red and then to green.

This is a reversible reaction: The solution also contains glucose which acts as a reducing agent, reducing the oxidised indigo carmine (GREEN) back to the reduced form (AMBER). This occurs as you let the solution stand.

Iron is Used to Make Steel

Iron is used in construction e.g. the Forth Bridge.

Iron is often mixed with other metals such as chromium, manganese, and nickel to give STEEL. Stainless steel is used in medicine to fix broken bones. It is strong and does not react inside the body (unlike iron which would rust).

8

Solid State Chemistry

SOLIDS IN LIQUIDS

Objectives

The first object was to find the density of various solid objects, both regular and irregular solids. The density of two liquids was to be determined. The second objective was to complete a full lab writeup as a model for the rest of the year.

- *Equipment*: Lab scales, graduated cylinder, vernier calipe
- *Safety rules*: Do Not Put the Sphere in the Graduated Cylinder!
- *Procedure*: See notecard
- *Data*: See datasheet from lab
- *Calculations*:
  - *Sphere 1*:

    $V = 4/3\pi\ r^3$ $\quad$ $(4/3)\ (\pi)\ (1.23\ \text{cm})^3 = 7.79\ \text{cm}^3$

    $\rho = m/V$ $\quad$ $2.8\ \text{g}/7.79\ \text{cm}^3 = 0.36\ \text{g/cm}^3$
  - *Sphere 2*:

    $V = 4/3\ \pi\ r^3$ $\quad$ $(4/3)\ (\pi)\ (1.79\ \text{cm})^3 = 24.0\ \text{cm}^3$

    $\rho = m/V$ $\quad$ $2.8\ \text{g}/24.0\ \text{cm}^3 = 0.12\ \text{g/cm}^3$
  - *Cylinder (metal)*:

    $V = \pi\ r^2\ L$ $\quad$ $(\pi)\ (1.35\ \text{cm})^2\ (5.10\ \text{cm}) = 29.2\ \text{cm}^3$

    $\rho = m/V$ $\quad$ $57.6\ \text{g}/29.2\ \text{cm}^3 = 1.96\ \text{g/cm}^3$
  - *Cylinder (metal using water displacement)*:

    $\rho = m/V$ $\quad$ $57.6\ \text{g}/28.1\ \text{mL} = 2.0\ \text{g/mL}$
  - *Rock*:

    $\rho = m/v$ $\quad$ $4.5\ \text{g}/2.3\ \text{mL} = 1.9\ \text{g/mL}$

– *Styrofoam block*:

$V = L \times W \times H$ (7.85 cm) (3.45 cm) (1.23 cm) = 33.3 cm^3

$\rho = m/V$ 1.3 g/33.3 cm^3 = 0.039 g/cm^3

– *Small cylinder*:

$V = \pi r^2 L$ (π) $(1.23\ cm)^2$ (3.46 cm) = 16.4 cm^3

$\rho = m/V$ 3.4 g/16.4 cm^3 = 0.21 g/cm^3

– *% Difference (metal cylinder)*:

% difference = | difference in 2 densities | /larger of 2 × 100

| 2.0 g/mL – 1.96 g/cm^3 |/2.0 g/mL × 100 = 2.0%

– *% error (metal cylinder)*:

% error = | actual value – experimental value /actual value × 100

| 1.96 g/cm^3 – 1.8 g/cm^3 |/1.8 g/cm^3 × 100 = 8.8%

Volume of water	*Mass of water*	*Volume of TDL*	*Mass of TDL*
10.1 mL	10.0 g	10.8 mL	14.7 g
21.3 mL	21.0 g	22.0 mL	30.0 g
30.7 mL	31.0 g	31.0 mL	42.2 g
42.0 mL	42.3 g	43.2 mL	58.7 g
50.5 mL	50.1 g	51.2 mL	70.8 g
14.456 mL	15.9 g	14.456 mL	19.7 g
76.00 mL	76.8 g	76.00 mL	103.4 g

- Density of water (from slope) = 1.01 g/mL
- Density of TDL (from slope) = 1.37 g/mL

Results:

- Density of sphere 1 = 0.36 g/cm^3
- Density of sphere 2 = 0.12 g/cm^3
- Density of metal cylinder (geometric) = 1.96 g/cm^3
- Density of metal cylinder (water displ) = 2.0 g/mL
- Density of rock = 1.9 g/mL
- Density of Styrofoam block = 0.039 g/cm^3
- Density of small cylinder = 0.21 g/cm^3
- % difference (metal cylinder) = 2.0%
- % error (metal cylinder) = 8.8%
- *Density of water*: 1.01 g/mL
- *Density of TDL*: 1.37 g/mL

The item with the greatest density (in this experiment) was the metal cylinder with a density of 1.96 g/cm^3. The item with the least density was the Styrofoam block with a density of 0.039 g/cm^3.

The TDL liquid was more dense than the water, as seen by a density of 1.3 g/mL compared to 1.01 g/mL for the water. Both spheres ended up with densities less than that of water, thus both would float in water.

The rock was more dense than water but slightly less dense than the metal cylinder. The% difference in densities between the metal cylinder (volume found by using a geometric formula) versus the density (volume found by water displacement) was low at 2.0%.

However, the 8.8% difference between the actual density of the metal cylinder and the correct value was too high (perhaps due to measurement error). Generalizations reached from this lab would include that metals tend to have higher densities than plastics, the shape of an object does not affect its density, and that Styrofoam is a material of exceedingly low density.

The lab techniques used in this lab would allow a person to identify unknown materials. The material's density would be obtained using the procedures used here and then checked with listings of known materials to help identify the material. The relationship between the dependent and independent variables would be stated as: as the volume increased (independent variable) the mass increased (dependent variable).

Discussion

Knowledge gained in doing this lab includes proper usage of the vernier calipers, graduated cylinders, lab scales, and lab notecards. No differences between prediction and result were noted.

Observations include bubbles forming in the TDL if it is added too rapidly, the difficulty of removing the TDL from the inside of the graduated cylinder, and the lack of uniformity (uniform shape) of the Styrofoam block.

Error could have been reduced in this lab by using the 0.01 gram lab scales instead of the 0.1 scales, confirming the readings on the vernier calipers with the teacher, and using more trials in the density of liquids lab (resulting in better line of best fit).

No questions regarding this lab were noted:

- Using the overflow method – fill a large container to the brim with water, add the object, capture and measure volume of overflow water (that volume = volume of object).
- The apparent volume would be greater than it should be and the density less than it should be.
- Use kitchen oil instead of water (in doing water displacement)
- To dive, large tanks are filled with water increasing density, to rise the water is pumped out.
- 5. $V = m/\rho$ 2.00 E 3 g/0.92 g/cm^3 = 2.2 E 3 cm^3

- Use a liquid (in place of water in water displacement) that have a density greater than that of the object (so the object sinks and displaces the liquid).
- Combine the massiveness of the particles with their compactness to find density.

Method

A body is weighed in air and then immersed in a liquid. The apparent loss in weight of the body when immersed in the liquid is, by Archimedes' principle, equal to the weight of liquid displaced by the body. From these measurements the density and specific gravity of either the solid body or the liquid may be determined.

Theory

The density of a substance is defined as its mass per unit volume and in the metric system of units is measured in grams per cubic centimetre. The specific gravity of a substance is defined as the ratio of the mass or weight of the substance to the mass or weight of an equal volume of water and is a pure number having zero dimensions. A convenient method for determining densities or specific gravities is one which uses the principle of Archimedes, namely, that when a body is immersed in a fluid there is exerted on the body a vertical upward force equal to the weight of fluid displaced.

Proof of Archimedes' Principle

Suppose aright circular cylinder of height h is immersed in a liquid of density d gm per cm^3. The bottom of the cylinder is at a depth of h_2 cm and the top at a depth h cm below the surface of the liquid where $h = (h_2 - h_1)$. The pressure on the liquid at depth h_1 is $p_1 = B + h_1gd$ dynes per cm^2, where B is atmospheric pressure.

The total downward force on the upper surface of the cylinder is $F_1 = p_1A$ dynes where A is the area of cross section of the cylinder. Similarly the total upward force on the bottom of the cylinder is $F_2 = p_2A$ dynes.

Since the pressure in a fluid acts at right angles to any surface in contact with it, the pressure on the sides of the cylinder is everywhere horizontal and has no vertical component. Hence the total upward force in dynes on the cylinder is,

$$F_2 - F_1 = p_2A = (h_2 - h_1)Agd = hAgd$$

But hA = V, the volume of the cylinder, so that,

$$F_2 - F_1 = Vgd$$

or the resultant upward force in dynes is equal to the weight of fluid displaced.While this proof of Archimedes' principle is for a particularly simple geometrical object, the principle is true for an object of any shape which is immersed in a fluid.

Application of Archimedes' Principle

- Suppose a body has a density of D grams per cm^3 and amass of M grams. The volume of the body is V = M/D and its weight in air is Mg dynes. The weight (MLg) of the body when immersed in a liquid of density d is, by Archimedes' principle, MLg = Mg – Vdg dynes. Thus ML = M – Vd or ML = M – Md/D. Since ML and M may be measured by means of a balance, it follows that if the density d of the liquid-say water-is known, the density D of the body may be calculated from,

$$D = Md/(M - M_L)$$

•

If the body whose density is to be measured is less dense than the liquid, it is necessary to fasten the body to a sinker so that the two together sink in the liquid. Let M_1g be the weight in air of the body, say a block of wood, whose density D_1 is less than that of the liquid of density d. Suppose the body of mass M and density D, referred to above, is such that when fastened to the block of wood the two together sink in the liquid. Let the weight of the two together when completely immersed in the liquid be MLg. Then from Figure. it is readily seen that,

$$M'_Lg = Mg + M_1g - (Mgd/D) - (M_1gd/D_1)$$

or

$$M'_L = M_L + M_1 - ((M_1d)/D_1)$$

Hence,

$$D_1 = (M_1d)/(M_L + M_1 - M'_L)$$

- The relative densities of liquids may be obtained by finding the weight of a sinker in the various liquids. Suppose that when the sinker is immersed in a liquid of density dl its apparent loss in weight is M_1g and when immersed in liquid of density d_2 the apparent loss in weight is M_2g. By Archimedes' principle $M_1g = Vd_1g$ and $M_2g = Vd_2g$ where V is the volume of the sinker.

Hence,

$$V = M_1/d_1 = M_2/d_2$$

or,

$$M_1/M_2 = d_1/d_2$$

M_1 and M_2 may be measured directly by a balance and if one of the liquids is water at a known temperature, its density may be found accurately in tables so that d_2 may be calculated.

Part I Apparatus

A balance of the type shown in Fig. is convenient for determining the mass of a body in air or in a liquid.

The two bodies whose densities are to be determined are, for example, a block of brass and a block of wood.

A suitable set of weights, a beaker, and a thin piece of cord for tying the two bodies together are necessary.

Procedure

By means of the balance determine the mass M of the block of brass in air.

Place a beaker containing water on the adjustable platform and suspend the brass block by means of a fine cord or wire from the hook from which the pan is suspended.

Completely immerse the brass block in the water. See that no bubbles adhere to the surface of the block. Determine the mass ML of the brass block in water.

Determine the mass of the block of wood in air. Fasten the wooden and brass blocks together by the cord or fine wire and determine the masses of the two together M'_L when completely immersed in water.

The density of water at the particular temperature may be obtained from Table. If time permits, determine the density d of alcohol or an oil by weighing the brass block in air and in the liquid. Calculate the density of the brass and wood.

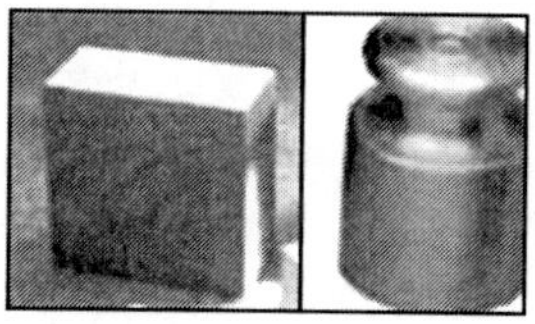

Fig. Blocks of Brass and Wood Suitable for Density Measurements.

Part II Apparatus

A Westphal balance which consists of a specially designed balance for measuring the specific gravity of liquids, is needed. A glass plummet, which is immersed in the liquid whose density or specific gravity is required, is suspended from the end of the calibrated arm of the balance.

The apparent loss of weight of the plummet when immersed in the liquid is measured by placing known weights on the arm until a balance is obtained. Four different weights having relative values of 1.0, 0.1, 0.01 and 0.001 are supplied. The glass plummet is constructed so that when immersed in water at 15.5°C the largest or unit weight placed on the hook with the glass plummet produces a balance.

Procedure

Place the leveling screw on the base of the instrument so that it is in the position under the arm of the balance. Clean and dry the glass plummet and hang it by the fine wire from the hook at the end of the arm of the balance. By means of the leveling screw adjust the arm of the balance so that it is horizontal- that is, so that the two pointers are in line.

Clean the glass hydrometre jar and partially fill it with distilled water. Immerse the glass plummet in the water and adjust the height of the instrument so that a single strand of the fine wire cuts the surface of the water. See that no bubbles adhere to the glass plummet. If the water is at 15.5°C, the unit weight hung at the end of the arm restores the balance. If the glass plummet is placed in other liquids at 15.5°C, the weights required to restore the balance directly measure the specific gravity of the liquids. For high accuracy in measuring the specific gravity or relative densities of liquids it is necessary to have the liquids at the same temperature. This is true not only because the density of liquids changes with temperature but also because the glass plummet changes in volume. If the temperatures are within a few degrees, the error introduced is not large.

Suppose that a balance is obtained when the glass plummet is immersed in water at 20°C with weights of 1.0 and 0.1 at notch 9, the 0.01 weight at notch 8 and the 0.001 weight at notch 5, and when immersed in another liquid at the same temperature a balance is obtained with weight 1.0 at notch 8, weights 0.1 and 0.01 at notch 6, and weight 0.001 at notch 4.

The relative densities of the two liquids are, by Eq.

$$d/d\omega = 0.8664/0.9985$$

The density of pure water at 20°C is given in Table I. Find the density of the various liquids provided, being careful to see that the glass plummet and the hydrometre jar are carefully cleaned each time before anew liquid is used. Read and record the temperature of the thermometre in the glass plummet each time.

Part III Apparatus

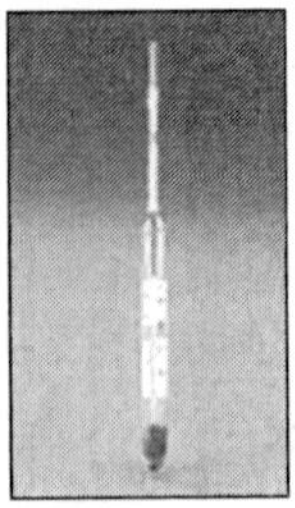

Fig. Hydrometre for Measuring Specific Gravities of Light Liquids.

A hydrometre of the type shown in Figure and a large hydrometre jar are required.

Procedure

Partially fill the hydrometre jar with the liquid whose specific gravity is to be measured and place the hydrometre in it. Clean the hydrometre and jar each time a new liquid is used. In this case the specific gravity of the liquid is read directly from the scale reading on the hydrometre. This method is not capable of giving as high an accuracy as that obtained with the Westphal balance, but the readings may be taken much more rapidly. Determine the specific gravities of the liquids used in the previous experiment.

CRYSTAL STRUCTURE

In mineralogy and crystallography, a crystal structure is a unique arrangement of atoms in a crystal. A crystal structure is composed of a motif, a set of atoms arranged in a particular way, and a lattice. Motifs are located upon the points of a lattice, which is an array of points repeating periodically in three dimensions. The points can be thought of as forming identical tiny boxes, called unit cells, that fill the space of the lattice. The lengths of the edges of a unit cell and the angles between them are called the lattice parameters. The symmetry properties of the crystal are embodied in its space group. A crystal's structure and symmetry play a role in determining many of its properties, such as cleavage, electronic band structure, and optical properties.

Crystallography and Crystal Defects

A plane of atoms can glide rigidly over its neighbour in process described as slip with the slip system defined by the plane and the direction of slip. This kind of deformation requires enormous stresses, far greater than those required to actually deform a crystal. This is because almost all crystals contain defects known as dislocations. A dislocation enables the planes to glide in a piecewise manner rather than the rigid displacement of the entire plane. This greatly reduces the stress required to cause slip.

A good analogy to illustrate the role of a dislocation is to imagine the force required to pull an entire carpet along the floor. On the other hand, if a bump is introduced into the carpet, the force needed to move the bump along is much smaller. Crystals may also contain point defects, which are imperfections of occupation. A vacancy is when an atom is missing from a site which should be occupied. An interstitial occurs when an atom is forced into a space within the crystal structure, where atoms are not normally located. Polycrystalline materials contain many crystals; another common term for crystals in such materials is grains. Atoms in the grain boundary between crystals must in general be displaced from positions they would occupy in the undisturbed crystal. Therefore, grain boundaries are defects.

Crystal Systems

The crystal systems are a grouping of crystal structures according to the axial system used to describe their lattice. Each crystal system consists of a set of three axes in a particular geometrical arrangement. There are seven unique crystal systems. The simplest and most symmetric, the cubic (or isometric) system, has the symmetry of a cube, that is, it exhibits four threefold rotational axes oriented at 109.5 degrees (the tetrahedral angle) with respect to each other. These threefold axes lie along the body diagonals of the cube. This definition of a cubic is correct, although many textbooks incorrectly state that a cube is defined by three mutually perpendicular axes of equal length – if this were true there would be far more than 14 Bravais lattices. The other six systems, in order of decreasing symmetry, are hexagonal, tetragonal, rhombohedral (also known as trigonal), orthorhombic, monoclinic and triclinic. Some crystallographers consider the hexagonal crystal system not to be its own crystal system, but instead a part of the trigonal crystal system. The crystal system and Bravais lattice of a crystal describe the (purely) translational symmetry of the crystal.

Point and Space Groups

The crystallographic point group or crystal class is the mathematical group comprising the symmetry operations that leave at least one point unmoved and that leave the appearance of the crystal structure unchanged. These symmetry operations can include reflection, which reflects the structure across a reflection plane, rotation, which rotates the structure a specified portion of a circle about a rotation axis, inversion which changes the sign of the coordinate of each point with respect to a centre of symmetry or inversion point and improper rotation, which consists of a rotation about an axis followed by an inversion. Rotation axes (proper and improper), reflection planes, and centers of symmetry are collectively called symmetry elements. There are 32 possible crystal classes. Each one can be classified into one of the seven crystal systems.

The space group of the crystal structure is composed of the translational symmetry operations in addition to the operations of the point group. These include pure translations which move a point along a vector, screw axes, which rotate a point around an axis while translating parallel to the axis, and glide planes, which reflect a point through a plane while translating it parallel to the plane. There are 230 distinct space groups.

PHYSICAL PROPERTIES

Defects or Impurities in Crystals

Real crystals feature defects or irregularities in the ideal arrangements described above and it is these defects that critically determine many of the electrical and mechanical properties of real materials. When one atom substitutes for one of the principal atomic components within the crystal structure, alteration in the electrical and thermal properties of the material may ensue. Impurities may also manifest as spin impurities in certain materials. Research on magnetic impurities demonstrates that substantial alteration of certain properties such as specific heat may be affected by small concentrations of an impurity, as for example impurities in semiconducting ferromagnetic alloys may lead to different properties as first predicted in the late 1960s. Dislocations in the crystal lattice allow shear at lower stress than that needed for a perfect crystal structure.

Crystal Symmetry and Physical Properties

Twenty of the 32 crystal classes are so-called piezoelectric, and crystals belonging to one of these classes (point groups) display piezoelectricity. All 21 piezoelectric classes lack a centre of symmetry. Any material develops a dielectric polarization when an electric field is applied, but a substance which has such a natural charge separation even in the absence of a field is called a polar material. Whether or not a material is polar is determined solely by its crystal structure. Only 10 of the 32 point groups are polar. All polar crystals are pyroelectric, so the 10 polar crystal classes are sometimes referred to as the pyroelectric classes.

There are a few crystal structures, notably the perovskite structure, which exhibit ferroelectric behaviour. This is analogous to ferromagnetism, in that, in the absence of an electric field during production, the ferroelectric crystal does not exhibit a polarisation. Upon the application of an electric field of sufficient magnitude, the crystal becomes permanently polarised. This polarisation can be reversed by a sufficiently large counter-charge, in the same way that a ferromagnet can be reversed.

However, it is important to note that, although they are called ferroelectrics, the effect is due to the crystal structure, not the presence of a ferrous metal. the angle between the normals to the two intersecting faces is called interfacial angle. Incommensurate crystals have period-varying translational symmetry. The period between nodes of symmetry is constant in most crystals. The distance between nodes in an incommensurate crystal is dependent on the number of nodes between it and the base node.

Interstices

The atoms inside a unit cell do not fill all space. The empty space represents the interstices. It is often the case that these interstices can accommodate small impurity atoms. As an example, we shall consider the crystal structure of iron which at ambient temperature has the cubic–I lattice with an atom of iron at each lattice point. There are two kinds of interstitial sites capable of accommodating small atoms such as carbon or nitrogen. These are the tetrahedral and octahedral sites as illustrated in.

Advantages of a Crystal

In principle, an atomic structure could be determined from applying X-ray scattering to non-crystalline samples, even to a single molecule. However, crystals offer a much stronger signal due to their periodicity. A crystalline sample is by definition periodic; a crystal is composed of many unit cells repeated indefinitely in three independent directions. Such periodic systems have a Fourier transform that is concentrated at periodically repeating points in reciprocal space known as Bragg peaks; the Bragg peaks correspond to the reflection spots observed in the diffraction image. Since the amplitude at these reflections grows linearly with the number N of scatterers, the observed intensity of these spots should grow quadratically, like N^2. In other words, using a crystal concentrates the weak scattering of the individual unit cells into a much more powerful, coherent reflection that can be observed above the noise. This is an example of constructive interference.

In a non-crystalline sample, molecules within that sample would be in random orientations and therefore would have a continuous Fourier spectrum that spreads its amplitude more uniformly and with a much reduced intensity, as is observed in SAXS.

More importantly, the orientational information is lost. In the crystal, the molecules adopt the same orientation within the crystal, whereas in a liquid, powder or amorphous state, the observed signal is averaged over the possible orientations of the molecules.

Although theoretically possible with sufficiently low-noise data, it is generally difficult to obtain atomic-resolution structures of complicated, asymmetric molecules from such rotationally averaged scattering data. An intermediate case is fibre diffraction in which the subunits are arranged periodically in at least one dimension.

Crystallization

Although crystallography can be used to characterize the disorder in an impure or irregular crystal, crystallography generally requires a pure crystal of high regularity to solve for the structure of a complicated arrangement of atoms. Pure, regular crystals can sometimes be obtained from Nature or man-made materials, such as samples of metals, minerals or other macroscopic materials.

The regularity of such crystals can sometimes be improved with annealing and other methods. However, in many cases, obtaining a diffraction-quality crystal is the chief barrier to solving its atomic-resolution structure. Small-molecule and macromolecular crystallography differ in the range of possible techniques used to produce diffraction-quality crystals.

Small molecules generally have few degrees of conformational freedom, and may be crystallized by a wide range of methods, such as chemical vapour deposition and recrystallisation. By contrast, macromolecules generally have many degrees of freedom and their crystallization must be carried out to maintain a stable structure. For example, proteins and larger RNA molecules cannot be crystallized if their tertiary structure has been unfolded; therefore, the range of crystallization conditions is restricted to solution conditions in which such molecules remain folded.

Protein crystals are almost always grown in solution. The most common approach is to lower the solubility of its component molecules very gradually; however, if this is done too quickly, the molecules will precipitate from solution, forming a useless dust or amorphous gel on the bottom of the container. Crystal growth in solution is characterized by two steps: nucleation of a microscopic crystallite (possibly having only 100 molecules), followed by growth of that crystallite, ideally to a diffraction-quality crystal. The solution conditions that favour the first step (nucleation) are not always the same conditions that favour the second step.

The crystallographer's goal is to identify solution conditions that favour the development of a single, large crystal, since larger crystals offer improved resolution of the molecule. Consequently, the solution conditions should disfavor the first step (nucleation) but favour the second (growth), so that only one large crystal forms per droplet. If nucleation is favoured too much, a shower of small crystallites will form in the droplet, rather than one large crystal; if favoured too little, no crystal will form whatsoever.

It is extremely difficult to predict good conditions for nucleation or growth of well-ordered crystals. In practice, favourable conditions are identified by screening; a very large batch of the molecules is prepared, and a wide variety of crystallization solutions are tested. Hundreds, even thousands, of solution conditions are generally tried before finding one that succeeds in crystallizing the molecules.

The various conditions can use one or more physical mechanisms to lower the solubility of the molecule; for example, some may change the pH, some contain salts of the Hofmeister series or chemicals that lower the dielectric constant of the solution, and still others contain large polymers such as polyethylene glycol that drive the molecule out of solution by entropic effects. It is also common to try several temperatures for encouraging crystallization, or to gradually lower the temperature so that the solution becomes supersaturated.

These methods require large amounts of the target molecule, as they use high concentration of the molecule(s) to be crystallized. Due to the difficulty in obtaining such large quantities (milligrams) of crystallisation grade protein, dispensing robots have been developed that are capable of accurately dispensing crystallisation trial drops that are of the order on 100 nanoliters in volume. This means that roughly 10-fold less protein is used per-experiment when compared to crystallisation trials setup by hand (on the order on 1 microliters), Several factors are known to inhibit or mar crystallization.

The growing crystals are generally held at a constant temperature and protected from shocks or vibrations that might disturb their crystallization. Impurities in the molecules or in the crystallization solutions are often inimical to crystallization. Conformational flexibility in the molecule also tends to make crystallization less likely, due to entropy. Ironically, molecules that tend to self-assemble into regular helices are often unwilling to assemble into crystals.

Crystals can be marred by twinning, which can occur when a unit cell can pack equally favourably in multiple orientations; although recent advances in computational methods have begun to allow the structures of twinned crystals to be solved, it is still very difficult. Having failed to crystallize a target molecule, a crystallographer may try again with a slightly modified version of the molecule; even small changes in molecular properties can lead to large differences in crystallization behaviour.

Mounting the Crystal

Animation showing the five motions possible with a four-circle kappa goniometer. The rotations about each of the four angles φ, κ, ω and 2θ are shown first. These rotations leave translucent crystal within the X-ray beam, but change the crystal orientation, allowing more of reciprocal space to be observed. Finally, the detector can be slid closer or further away from the crystal, allowing higher resolution data to be taken (if closer) or better discernment of the Bragg peaks.

Once they are full-grown, the crystals are mounted so that they may be held in the X-ray beam and rotated. There are several methods of mounting. Although crystals were once loaded into glass capillaries with the crystallization solution (the mother liquor), a more modern approach is to scoop the crystal up in a tiny loop, made of nylon or plastic and attached to a solid rod, that is then flash-frozen with liquid nitrogen. This freezing reduces the radiation damage of the X-rays, as well as the noise in the Bragg peaks due to thermal motion (the Debye-Waller effect). However, untreated crystals often crack if flash-frozen; therefore, they are generally pre-soaked in a cryoprotectant solution before freezing. Unfortunately, this pre-soak may itself cause the crystal to crack, ruining it for crystallography. Generally, successful cryo-conditions are identified by trial and error.

The capillary or loop is mounted on a goniometer, which allows it to be positioned accurately within the X-ray beam and rotated. Since both the crystal and the beam are often very small, the crystal must be centred within the beam to within roughly 25 micrometres accuracy, which is aided by a camera focused on the crystal. The most common type of goniometer is the "kappa goniometer", which offers three angles of rotation: the ω angle, which rotates about an axis roughly perpendicular to the beam; the κ angle, about an axis at roughly 50° to the ω axis; and, finally, the φ angle about the loop/capillary axis. When the κ angle is zero, the ω and φ axes are aligned. The κ rotation allows for convenient mounting of the crystal, since the arm in which the crystal is mounted may be swung out towards the crystallographer. The oscillations carried out during data collection involve the ω axis only. An older type of goniometer is the four-circle goniometer, and its relatives such as the six-circle goniometer.

X-ray Sources

The mounted crystal is then irradiated with a beam of monochromatic X-rays. The brightest and most useful X-ray sources are synchrotrons; their much higher luminosity allows for better resolution. They also make it convenient to tune the wavelength of the radiation, which is useful for multi-wavelength anomalous dispersion (MAD) phasing. Synchrotrons are generally national facilities, each with several dedicated beamlines where data is collected around the clock, seven days a week.

Smaller, weaker X-ray sources are often used in laboratories to check the quality of crystals before bringing them to a synchrotron and sometimes to solve a crystal structure. In such systems, electrons are boiled off of a cathode and accelerated through a strong electric potential of roughly 50 kV; having reached a high speed, the electrons collide with a metal plate, emitting bremsstrahlung and some strong spectral lines corresponding to the excitation of inner-shell electrons of the metal. The most common metal used is copper, which can be kept cool easily, due to its high thermal conductivity, and which produces strong $K\alpha$ and $K\beta$ lines. The $K\beta$ line is sometimes suppressed with a thin layer (0.0005 in. thick) of nickel foil. The simplest and cheapest variety of sealed X-ray tube has a stationary anode (the Crookes tube) and produces circa 2 kW of X-ray radiation. The more expensive variety has a rotating-anode type source that produces circa 14 kW of X-ray radiation.

X-rays are generally filtered to a single wavelength (made monochromatic) and collimated to a single direction before they are allowed to strike the crystal. The filtering not only simplifies the data analysis, but also removes radiation that degrades the crystal without contributing useful information. Collimation is done either with a collimator (basically, a long tube) or with a clever arrangement of gently curved mirrors. Mirror systems are preferred for small crystals (under 0.3 mm) or with large unit cells (over 150 Å)

Solids State Chemistry

Amorphous solids are homogeneous and isotropic because there is no long range order or periodicity in their internal atomic arrangement. By contrast, the crystalline state is characterised by a regular arrangement of atoms over large distances. Crystals are therefore anisotropic – their properties vary with direction. For example, the interatomic spacing varies with orientation within the crystal, as does the elastic response to an applied stress. Engineering materials are usually aggregates of many crystals of varying sizes and shapes; these polycrystalline materials have properties which depend on the nature of the individual crystals, but also on aggregate properties such as the size and shape distributions of the crystals, and the orientation relationships between the individual crystals.

The randomness in the orientation of the crystals is a measure of texture, which has to be controlled in the manufacture of transformer steels, uranium fuel rods and beverage cans.

The crystallography of interfaces connecting adjacent crystals can determine the deformation behaviour of the polycrystalline aggregate; it can also influence the toughness through its effect on the degree of segregation of impurities to such interfaces.

Metals and ceramics are the basic materials of nuclear reactors. The regular arrangement of the atoms in such solids is termed a crystal lattice. Pure solids (whether elemental or binary) need not retain the same crystal structure from low temperature to the melting point; iron and uranium exhibit three crystal structures, each confined to a definite temperature interval, but uranium dioxide has only one. The change from one crystal structure to another is called a phase transformation. Such changes are the solid-solid equivalents of melting and vaporization.

The crystal structure can exert a profound effect on material's behaviour in the environment of a nuclear reactor, as seen from the following examples:

- Void swelling can cause unacceptable dimensional changes in structures of fast reactors or fusion reactors (but not generally in light water reactors, except at higher doses and intermediate temperatures in austenitic stainless steels). The crystal structure of the austenitic form of steel is susceptible to this radiation effect but the different crystal structure of the ferritic form is immune until very high doses (>H"50 displacements per atom, dpa).
- Under fast-neutron irradiation, Zircaloy cladding of LWRs undergoes growth in the axial direction. This effect is due to the anisotropy of the crystal structure of zirconium.
- The relation between crystal anisotropy and the directional dependence of macroscopic properties in zirconium was covered in Sect I.2.
- The utility of uranium metal as a nuclear fuel is diminished because of the volume change accompanying a solid phase transformation that occurs at relatively low temperature. Imperfections in the crystal lattice are even more important than the details of the atomic structure.
- Missing atoms in the atomic lattice called vacancies, or extra atoms in nonregular positions called interstitials, are responsible for the mobility of atoms in the solid.
- Many of the mechanical properties of metals are governed by a linear defect in the atomic structure called a dislocation.
- All structural alloys and ceramic fuels consist of a large number of small crystals, or grains, joined at their peripheries. The discontinuities in the crystal structure at these internal surfaces are called grain boundaries. Their presence influences a variety of mechanical and transport properties, including cracking in steel and fission gas movement in fuel.

For elemental solids, only 14 distinct structures, or spatial arrangements of atoms, are possible. Each crystal structure is characterized by a unit cell, which is a small group of atoms that contains the unique features of the particular lattice type. A macroscopic crystal is composed of large numbers of unit cells stacked together.

The elemental and binary solids important in LWRs are restricted to only a few of the 14 possible lattice types, mainly those with the highest degrees of symmetry and the tightest packing of atoms. With one exception, these unit cells are parallelepipeds (all angles 90o) with side lengths denoted by a, b, and c.

The exception is a structure whose unit cell is a hexagonal prism with only two characteristic dimensions. Only the parallelepiped with all sides of equal length (*i.e.*, a cube) is isotropic. All the others are anisotropic, meaning that properties differ in the principal coordinate directions.

CRYSTALLOGRAPHY

Crystal Structure

In mineralogy and crystallography, a crystal structure is a unique arrangement of atoms in a crystal. A crystal structure is composed of a motif, a set of atoms arranged in a particular way, and a lattice. Motifs are located upon the points of a lattice, which is an array of points repeating periodically in three dimensions. The points can be thought of as forming identical tiny boxes, called unit cells, that fill the space of the lattice. The lengths of the edges of a unit cell and the angles between them are called the lattice parameters. The symmetry properties of the crystal are embodied in its space group. A crystal's structure and symmetry play a role in determining many of its properties, such as cleavage, electronic band structure, and optical properties.

Solids

Amorphous solids are homogeneous and isotropic because there is no long range order or periodicity in their internal atomic arrangement. By contrast, the crystalline state is characterised by a regular arrangement of atoms over large distances. Crystals are therefore anisotropic – their properties vary with direction. For example, the interatomic spacing varies with orientation within the crystal, as does the elastic response to an applied stress. Engineering materials are usually aggregates of many crystals of varying sizes and shapes; these polycrystalline materials have properties which depend on the nature of the individual crystals, but also on aggregate properties such as the size and shape distributions of the crystals, and the orientation relationships between the individual crystals.

The randomness in the orientation of the crystals is a measure of texture, which has to be controlled in the manufacture of transformer steels, uranium fuel rods and beverage cans. The crystallography of interfaces connecting adjacent crystals can determine the deformation behaviour of the polycrystalline aggregate; it can also influence the toughness through its effect on the degree of segregation of impurities to such interfaces.

Metals and ceramics are the basic materials of nuclear reactors. The regular arrangement of the atoms in such solids is termed a crystal lattice. Pure solids (whether elemental or binary) need not retain the same crystal structure from low temperature to the melting point; iron and uranium exhibit three crystal structures, each confined to a definite temperature interval, but uranium dioxide has only one. The change from one crystal structure to another is called a phase transformation. Such changes are the solid-solid equivalents of melting and vaporization.

The crystal structure can exert a profound effect on material's behaviour in the environment of a nuclear reactor, as seen from the following examples:

- Void swelling can cause unacceptable dimensional changes in structures of fast reactors or fusion reactors (but not generally in light water reactors, except at higher doses and intermediate temperatures in austenitic stainless steels). The crystal structure of the austenitic form of steel is susceptible to this radiation effect but the different crystal structure of the ferritic form is immune until very high doses (>H"50 displacements per atom, dpa).
- Under fast-neutron irradiation, Zircaloy cladding of LWRs undergoes growth in the axial direction. This effect is due to the anisotropy of the crystal structure of zirconium.
- The relation between crystal anisotropy and the directional dependence of macroscopic properties in zirconium was covered in Sect I.2.
- The utility of uranium metal as a nuclear fuel is diminished because of the volume change accompanying a solid phase transformation that occurs at relatively low temperature.Imperfections in the crystal lattice are even more important than the details of the atomic structure.
- Missing atoms in the atomic lattice called vacancies, or extra atoms in nonregular positions called interstitials, are responsible for the mobility of atoms in the solid.
- Many of the mechanical properties of metals are governed by a linear defect in the atomic structure called a dislocation.
- All structural alloys and ceramic fuels consist of a large number of small crystals, or grains, joined at their peripheries. The discontinuities in the crystal structure at these internal surfaces are called grain boundaries. Their presence influences a variety of mechanical and transport properties, including cracking in steel and fission gas movement in fuel.

For elemental solids, only 14 distinct structures, or spatial arrangements of atoms, are possible. Each crystal structure is characterized by a unit cell, which is a small group of atoms that contains the unique features of the particular lattice type. A macroscopic crystal is composed of large numbers of unit cells stacked together.

The elemental and binary solids important in LWRs are restricted to only a few of the 14 possible lattice types, mainly those with the highest degrees of symmetry and the tightest packing of atoms. With one exception, these unit cells are parallelepipeds (all angles 90o) with side lengths denoted by a, b, and c.

The exception is a structure whose unit cell is a hexagonal prism with only two characteristic dimensions. Only the parallelepiped with all sides of equal length (*i.e.*, a cube) is isotropic. All the others are anisotropic, meaning that properties differ in the principal coordinate directions.

The Lattice

Crystals have translational symmetry: it is possible to identify a regular set of points, known as the lattice points, each of which has an identical environment. The set of these lattice points constitutes a three dimensional lattice. A unit cell may be defined within this lattice as a space–filling parallelepiped with origin at a lattice point, and with its edges defined by three non-coplanar basis vectors a1, a2 and a3, each of which represents translations between two lattice points.

The entire lattice can then be generated by stacking unit cells in three dimensions. Any vector representing a translation between lattice points is called a lattice vector. The unit cell defined above has lattice points located at its corners. Since these are shared with seven other unit cells, and since each cell has eight corners, there is only one lattice point per unit cell. Such a unit cell is primitive and has the lattice symbol P. Non–primitive unit cells can have two or more lattice points, in which case, the additional lattice points will be located at positions other than the corners of the cell. A cell with lattice points located at the centres of all its faces has the lattice symbol F; such a cell would contain four lattice points. Not all the faces of the cell need to have face–centering lattice points; when a cell containing two lattice points has the additional lattice point located at the centre of the face defined by a2 and a3, the lattice symbol is A and the cell is said to be A-centred. B-centred and C-centred cells have the additional lattice point located on the face defined by a3 & a1 or a1 & a2 respectively.

A unit cell with two lattice points can alternatively have the additional lattice point at the body-centre of the cell, in which case the lattice symbol is I. The lattice symbol R is for a trigonal cell; the cell is usually defined such that it contains three lattice points. The basis vectors a1, a2 and a3 define the the unit cell; their magnitudes a1, a2 and a3 respectively, are the lattice parameters of the unit cell. The angles a1^a2, a2^a3 and a3 ^a1 are conventionally labelled g,aand b respectively. Note that our initial choice of the basis vectors was arbitrary since there are an infinite number of lattice vectors which could have been used in defining the unit cell. The preferred choice includes small basis vectors which are as equal as possible, provided the shape of the cell reflects the essential symmetry of the lattice.

The Bravais Lattices

The number of ways in which points can be arranged regularly in three dimensions, such that the stacking of unit cells fills space, is not limitless;

Bravais showed in 1848 that all possible arrangements can be represented by just fourteen lattices. The fourteen Bravais lattices can be categorised into seven crystal systems (cubic, tetragonal, orthorhombic, trigonal, hexagonal, monoclinic and triclinic); the cubic system contains for example, the cubic-P, cubic-F and cubic-I lattices. Each crystal system can be characterised uniquely by a set of defining symmetry elements, which any crystal within that system must possess as a minimum requirement.

The Bravais lattices are each as a projection along the a_3 axis. Projections like these are useful as simple representations of three–dimensional objects. The coordinate of any point with respect to the a_3 axis is represented as a fraction of a_3 along the point of interest; points located at $0a3$ are unlabelled; translational symmetry requires that for each lattice point located at $0a_3$, there must exist another at $1a_3$.

Directions

Any vector u can be represented as a linear combination of the basis vectors ai of the unit cell (i = 1, 2, 3):

$$u = u_1a_1 + u_2a_2 + u_3a_3$$

and the scalar quantities u_1, u_2 and u_3 are the components of the vector u with respect to the basis vectors a_1, a_2 and a_3. Once the unit cell is defined, any direction u within the lattice can be identified uniquely by its components [u1 u2 u3], and the components are called the Miller indices of that direction and are by convention enclosed in square brackets.

Planes

If a plane intersects the a1, a2 and a3 axes at distances x1, x2 and x3 respectively, relative to the origin, then the Miller indices of that plane are given by (h1 h2 h3) where:

$$h_1 = \phi\frac{a_1}{x_1},\quad h_2 = \phi\frac{a_2}{x_2},\quad h_3 = \phi\frac{a_3}{x_3},$$

ϕ is a scalar which clears the numbers *hi* off fractions or common factors. Note that *xi* are negative when measured in the "*ai* directions. The intercept of the plane with an axis may occur at 1, in which case the plane is parallel to that axis and the corresponding Miller index will be zero. Miller indices for planes are by convention written using round brackets: (h1 h2 h3) with braces being used to indicate planes of the same form: {h1 h2 h3}.

The Reciprocal Lattice

The reciprocal lattice is a special co–ordinate system. For a lattice represented by basis vectors a1, a2 and a3, the corresponding reciprocal basis vectors are written a_1^*, a_2^* and a_3^*, such that:

$$a_1^* = \frac{(a_2 \wedge a_3)}{(a_1.a_2 \wedge a_3)}$$

$$a_2^* = \frac{(a_3 \wedge a_1)}{(a_1.a_2 \wedge a_3)}$$

$$a_3^* = \frac{(a_1 \wedge a_2)}{(a_1.a_2 \wedge a_3)}$$

In equation 2a, the term $(a_1.a_2 \wedge a_3)$ represents the volume of the unit cell formed by ai, while the magnitude of the vector $(a_2 \wedge a3)$ represents the area of the (1 0 0)A plane. Since $(a_2 \wedge a_3)$ points along the normal to the (1 0 0)A plane, it follows that a_1^* also points along the normal to (1 0 0)A and that its magnitude | | is the reciprocal of the spacing of the (1 0 0)A planes.

The components of any vector referred to the reciprocal basis represent the Miller indices of a plane whose normal is along that vector, with the spacing of the plane given by the inverse of the magnitude of that vector. For example, the reciprocal lattice vector u* = (1 2 3) is normal to planes with Miller indices (1 2 3) and interplanar spacing 1/|u*|.

$$a_i \cdot a_j^* = 1 \quad \text{when } i = j, \quad \text{and } ai \cdot a_j^* = 0 \quad \text{when } i \neq j$$

or in other words, $a_i \cdot a_j^* = \delta_{ij}$.

δ_{ij} is the Kronecker delta, which has a value of unity when i = j and is zero when $i \neq j$

Electrons in Metals

Metals are electrical conductors because some of the electrons are able to move freely, even though they move in a periodic array of positive metal ions. Electrons are waves, characterised by a wave number *k*,

$$k = \pm\frac{2\pi}{\lambda}$$

where λ is the wavelength. The lattice does not affect the motion of these electrons except at critical values of k, where the lattice planes reflect the electrons†. For a square lattice, these critical values of *k* are given by,

$$k = \frac{n\pi}{a \sin\theta}$$

where a is the lattice parameter and μ is the direction of motion. The component of k along the *x* axis is $kx = k \sin \mu$ and since μ = 0, it follows that reflection occurs when

$$kx = \pm\frac{1/4}{a}$$

and similarly when,

$$ky = \pm\frac{1/4}{a},$$

This can be represented in *k*–space, which is reciprocal space with a reciprocal lattice parameter of magnitude 2p/a. Electrons are reflected in the lattice whenever k falls on any point on the square ABCD. This square is known as the first Brillouin zone boundary. Electron energy contours are frequently plotted in *k*–space. Inside a Brillouin zone, the energy of electrons can increase 'smoothly' with *k*; there is then a discontinuity (an energy gap) at the zone boundary.

Unit Cell

The crystal structure of a material or the arrangement of atoms in a crystal structure can be described in terms of its unit cell. The unit cell is a tiny box containing one or more motifs, a spatial arrangement of atoms. The units cells stacked in three-dimensional space describes the bulk arrangement of atoms of the crystal. The unit cell is given by its lattice parameters, the length of the cell edges and the angles between them, while the positions of the atoms inside the unit cell are described by the set of atomic positions (*xi, yi, zi*) measured from a lattice point.

Although there are an infinite number of ways to specify a unit cell, for each crystal structure there is a conventional unit cell, which is chosen to display the full symmetry of the crystal. However, the conventional unit cell is not always the smallest possible choice. A primitive unit cell of a particular crystal structure is the smallest possible volume one can construct with the arrangement of atoms in the crystal such that, when stacked, completely fills the space. This primitive unit cell will not always display all the symmetries inherent in the crystal. A Wigner-Seitz cell is a particular kind of primitive cell which has the same symmetry as the lattice. In a unit cell each atom has an identical environment when stacked in 3 dimensional space. In a primitive cell, each atom may not have the same environment.

Symmetry

Although the properties of a crystal can be anisotropic, there may be different directions along which they are identical. These directions are said to be equivalent and the crystal is said to possess symmetry. That a particular edge of a cube cannot be distinguished from any other is a measure of its symmetry; an orthorhombic parallelepiped has lower symmetry, since its edges can be distinguished by length. Some symmetry operations are illustrated in Fig. 6; in essence, they transform a spatial arrangement into another which is indistinguishable from the original. The rotation of a cubic lattice through

90° about an axis along the edge of the unit cell is an example of a symmetry operation, since the lattice points of the final and original lattice coincide in space and cannot consequently be distinguished. We have already encountered translational symmetry when defining the lattice; since the environment of each lattice point is identical, translation between lattice points has the effect of shifting the origin.

Symmetry Operations

An object possesses an n-fold axis of rotational symmetry if it coincides with itself upon rotation about the axis through an angle $360°\pm/n$. The possible angles of rotation, which are consistent with the translational symmetry of the lattice, are 360°, 180°, 120°, 90°and 60° for values of n equal to 1, 2, 3, 4 and 6 respectively. A five–fold axis of rotation does not preserve the translational symmetry of the lattice and is forbidden. A one-fold axis of rotation is called a monad and the terms diad, triad, tetrad and hexad correspond to n = 2, 3, 4 and 6 respectively.

All of the Bravais lattices have a centre of symmetry. An observer at the centre of symmetry sees no difference in arrangement between the directions $[u_1\ u_2\ u_3]$ and $[-u_1\ -u_2\ -u_3]$. The centre of symmetry is such that inversion through that point produces an identical arrangement but in the opposite sense. A rotoinversion axis of symmetry rotates a point through a specified angle and then inverts it through the centre of symmetry such that the arrangements before and after this combined operation are in coincidence. For example, a three-fold inversion axis involves a rotation through 120° combined with an inversion, the axis being labelled $\overline{3}$.

A rotation operation can also be combined with a translation parallel to that axis to generate a screw axis of symmetry. The magnitude of the translation is a fraction of the lattice repeat distance along the axis concerned. A 31 screw axis would rotate a point through 120° and translate it through a distance t/3, where t is the magnitude of the shortest lattice vector along the axis. A 32 operation involves a rotation through 120° followed by a translation through 2t/3 along the axis. For a right–handed screw axis, the sense of rotation is anticlockwise when the translation is along the positive direction of the axis. A plane of mirror symmetry implies arrangements which are mirror images. Our left and right hands are mirror images. The operation of a 2 axis produces a result which is equivalent to a reflection through a mirror plane normal to that axis. The operation of a glide plane combines a reflection with a translation parallel to the plane, through a distance which is half the lattice repeat in the direction concerned. The translation may be parallel to a unit cell edge, in which case the glide is axial; the term diagonal glide refers to translation along a face or body diagonal of the unit cell. In the latter case, the translation is through a distance which is half the length of the diagonal concerned, except for diamond glide, where it is a quarter of the diagonal length.

Classification of Crystals by Symmetry

The defining property of a crystal is its inherent symmetry, by which we mean that under certain operations the crystal remains unchanged. For example, rotating the crystal 180 degrees about a certain axis may result in an atomic configuration which is identical to the original configuration. The crystal is then said to have a twofold rotational symmetry about this axis. In addition to rotational symmetries like this, a crystal may have symmetries in the form of mirror planes and translational symmetries, and also the so-called compound symmetries which are a combination of translation and rotation/mirror symmetries. A full classification of a crystal is achieved when all of these inherent symmetries of the crystal are identified.

UNIT CELL LAWS OF CRYSTALLOGRAPHY

It is perhaps not generally appreciated to what an extent modern work on crystal structure is based upon the fundamental principles enunciated by Haüy over 125 years ago. The terminology may be different, but the ideas remain the same. He spoke of"integral molecules," we say"crystal molecules,""unit cells of the space lattices," etc. He adopted as the shapes of these units the parallelopiped, octahedron, tetrahedron, hexagonal prism, rhombic dodecahedron, and hexagonal bipyramid. We still use some of these, altho certain changes in the list have been required as a result of subsequent investigations. His law of rational indices has held firm, and has furnished a starting point for many subsequent generalizations.

The point-system or space-lattice theory of crystal structure is a direct extension of it. Professor Goldschmidt's recognition of harmony and complication in the symbols of zones of forms is another. And now that the use of X-rays has furnished, especially in the hands of the Braggs, a means of demonstrating the reality of space lattices and the occupation of their nodes by atoms, rather than by chemical molecules, it is possible to go still further.

The Russian crystallographer, E. S. Fedorov, has been during the past few, years carrying on extensive researches on what he terms"crystallo-chemistry," and, aided by the Braggs's working out of the structures of a number of simple salts, has recently formulated certain laws and principles extending Haüy's original law into this field.1 Being written in Russian,2his papers are comparatively inaccessible to English-speaking scientists, but it has fortunately been possible to have very full and accurate abstracts of them published in Chemical Abstracts.

The principal points made by Professor Fedorov are: The space arrangement of atoms in three crystallographic directions, and the units to be used in the corresponding axes, may be expressed rationally, altho, because of our inability to determine the positions of the atoms with absolute exactness, only approximately rational relations are as a rule obtained.

Planes determined by any three atoms are possible crystal faces, but the greater the density of a face the more important it is, crystallographically speaking, and the more likely to appear on the actual crystal.

The importance of planes is increased by the presence in them of dissimilar atoms, capable of attracting one another, and due to similar chemical attractions the bond between parallel planes is made stronger.

The nearest distance between unlike atoms in any space-lattice is in general inversely proportional to their chemical affinity, altho exceptions may occur when an atom of one kind holds two or more of another. Thruout these theorems the influence of Haüy's law of rational indices can be clearly traced.

An extension of the idea of arrangement of atoms in layers into the field of optical properties has recently been attempted by the writer. Thus far only tetragonal crystals have been studied, but in a number of substances which crystallize in this system there has been found to be a correlation between crystallographic axial ratio and refractive indices.

Deriving from the indices, by the use of the Lorentz-Lorenz formula, the"refractions," the following inverse relationship has been found to hold in certain simple compounds,

$$\frac{W2 - 1}{W2 + 2} : \frac{E^2 - 1}{E^2 + 2} = c : a.$$

It is of course necessary that the true axial ratio, based on all the layers of atoms present, be used; and in many cases this is not the same as the standard axial ratio, which is obtained by taking the most prominent pyramid form to be (111). There are, furthermore, many disturbing factors, such as irregularity of the layers, asymmetric arrangements of atoms, which produce rotation of the plane of polarized light, and the presence of absorption bands and other colour phenomena, such as pleochroism. Because of the presence of one or more of these features, the effects of which cannot as yet be evaluated, this rule fails to hold with complex compounds, but it does hold for enough simple ones to indicate that it is a fundamental relationship.

In a preliminary paper on this subject, recently published,[4] a few organic compounds were considered. Similar relations prove to hold with the minerals of the zircon group, the crystallographic features of which were first established by Haüy.

In cassiterite the refraction ratio is 0.945, the alternate axial ratio 0.951; in rutile the values are 0.926 and 0.911 respectively.

The practical identity of the two ratios shows that in these minerals the space-lattice must have the same number of layers of atoms in the horizontal as in the vertical direction. In zircon and xenotime the refraction ratios are not equal to either axial ratio, but to 3/2 the standard one.

This is interpreted to show that in the unit cells of these minerals there are three layers of atoms horizontally for every two vertically. It is accordingly possible to tell in this way something about the space-lattices of crystals, and when the refraction data are combined with the observed symmetry relations, to work out in detail probable arrangements of atoms for the substances.

A few minerals in addition to those of the zircon group have also been studied in this way, but in most cases the presence of disturbing factors prevents the complete recognition of their structure.

Further work is planned, however, to evaluate as far as possible the influence of these factors, as well as to apply this method of study to substances crystallizing in other systems than the tetragonal.

Crystallography

"The branch of science that deals with the study of structure, geometry and properties of crystals is called crystallography".

Symmetry in Crystal

A crystal possess following three types of symmetry,

Plane of Symmetry

It is an imaginary plane which passes through the centre of a crystal can divides it into two equal portions which are exactly the mirror images of each other.

Axis of Symmetry

An axis of symmetry or axis of rotation is an imaginary line, passing through the crystal such that when the crystal is rotated about this line, it presents the same appearance more than once in one complete revolution *i.e.,* in a rotation through 360°.

Suppose, the same appearance of crystal is repeated, on rotating it through an angle of 360°/n, around an imaginary axis, is called an n-fold axis where,n is known as the order of axis. By order is meant the value of n in $2\pi/n$ so that rotation through $2\pi/n$, gives an equivalent configuration.

Centre of Symmetry

It is an imaginary point in the crystal that any line drawn through it intersects the surface of the crystal at equal distance on either side.

Only simple cubic system have one centre of symmetry. Other system do not have centre of symmetry.

The total number of planes, axes and centre of symmetries possessed by a crystal is termed as elements of symmetry.

A cubic crystal possesses total 23 elements of symmetry:

- *Plane of symmetry*: (3 + 6) = 9
- *Axes of symmetry*: (3 + 4 + 6) = 13
- *Centre of symmetry*: (1) = 1
- (Total symmetry = 232) Laws of crystallography: Crystallography is based on three fundamental laws.

Law of Constancy of Interfacial Angles

This law states that angle between adjacent corresponding faces is inter facial angles of the crystal of a particular substance is always constant inspite of different shapes and sizes and mode of growth of crystal. The size and shape of crystal depend upon the conditions of crystallisation. This law is also known as Steno's Law.

Law of Rational Indices

This law states that the ratio of intercepts of different faces of a crystal with the three axes are constant and can be expressed by rational numbers that the intercepts of any face of a crystal along the crystallographic axes are either equal to unit intercepts (*i.e.*, intercepts made by unit cell) a,b, c or some simple whole number multiples of them *e.g.*, na, n' b, n'c, where n, n' and n' are simple whole numbers.

The whole numbers n, n' and n' are called Weiss indices. This law was given by Hauy.

LAW OF THE CONSTANCY OF INTERFACIAL ANGLES

Since the internal structure of any mineral is always constant, and since the possible crystal faces of that mineral have a definite relationship to that structure, it follows that mineral have a definite relationship to that structure, it follows that the faces must have also a definite relationship to each other.

This fact may be stated as follows: The angles between two similar faces on the same substance are always the same. Figure will also illustrate this point.

The face which cuts the network along the line A-C must make an angle of 45 degrees with the face which cuts along the line A-B, etc. This law is the most fundamental and important in the science of crystallography. It frequently enables one to identify a mineral by the measurement of the interfacial angles on its crystals. A mineral may be found in crystals of widely varying shapes and sizes, but the angle between two similar faces will always be the same.

And important part of the study of crystallography consists in the measuring and classifying of the interfacial angles on the crystals of all minerals. These measurements are accomplished by means of instruments know as goniometry.

For accurate work, particularly in the case of small crystals, a type of instrument known as a reflection goniometre is used. This is an instrument upon which the crystal to be measured is mounted so as to reflect beams of light from its faces through a telescope to the eye.

The size of the angle through which a crystal has to be turned in order to throw successive beams of light from two adjacent faces into tie telescope determines the angle existing between the faces. A simpler instrument used for approximate work and with larger crystals is known as a contact goniometre.

The regular internal structure of crystals requires that the ultimate individual mineral units must be at least physically alike. A physical likeness between these units necessitates that they should also be the same chemically, or at least closely similar.

Consequently we can state that in general a crystal must be made up of a regular assemblage of units which are chemically the same, and therefore that a crystallized mineral must have a definite and uniform chemical composition.

This statement is a general one and will suffice for the present; certain modifications will be found stated on under isomorphism. A crystal is a guaranty of the chemical homogeneity of a mineral. From this it follows that only definite chemical compounds are capable of crystallization.

To sum up the conclusions of the preceding paragraphs: A crystal is a solid with definite chemical composition which possesses a definite internal arrangement of its mineral units.

These internal characteristic are expressed outwardly in a definite external form. And since the internal structure of the same substance is always constant, the angles between the similar bounding planes of the crystals of that substance are also constant.

SOLIDS IN OPTICAL PROPERTIES

A particle placed in a beam of light will scatter some of the light incident on it (*i.e.* the light will change direction) and absorb some of the light (*i.e.* the electromagnetic energy is transformed into other forms of energy by the particle). The light we then receive has been attenuated or, to put it another way, suffered extinction. The amount of light scattered and/or absorbed by the particle depends, in an intricately involved way, on the exact nature of the particle, and also the nature of the incident light. Consequently the extinction is dependent on the chemical composition of the particle and its size, shape and orientation, together with the wavelength and polarization state of the light. Therefore, we must include all these factors in our interpretations of spectra.

Identification of astronomical dust is usually achieved by comparing the observed spectral data with a catalogue of laboratory spectra of materials considered to be similar to those found in the astronomical dust. A computer programme can be used to make correlations between the reference spectra and the observed spectra, however there are limitations to this technique. In order to compare the laboratory and observed spectra they need to be of similar type, *i.e.* transmission spectra of very fine particulate materials.

Many of the available spectra are obtained from samples for which no attempt has been made to simulate the possible conditions in the observed astronomical dust.

One of the reasons for this is that the laboratories want to optimize the surface conditions of the sample to yield the maximum amount of spectral information. Also, the applications to which infrared spectra are put are diverse and therefore the collection of such spectra uses many different techniques, producing different types of spectra, according to the application. A complete understanding of the transmission and reflection properties of both particulate and bulk media is imperative for the correct interpretation of the spectral data.

Physical Optics

Absorption, reflection, scattering, polarization and dispersion of light by a material medium can all be explained in terms of the atoms, molecules and lattice structures which make up the medium. This section will concentrate on the mechanisms by which electromagnetic radiation is absorbed by matter.Absorption of radiation by matter always involves the loss of energy by the radiation and a gain in energy by the atoms or molecules of the medium.

The energy of an assembly of atoms consists partly of thermal energy and partly of internal energy associated with the binding of the extra-nuclear electrons to the nucleus and with the binding of the particles within the nucleus itself. Molecules have, in addition, energy associated with the oscillations of the atoms within the molecule with respect to one another, and solids have energy associated with intermolecular bonds. The energy absorbed from radiation appears as increased internal energy or as increased vibrational energy of interatomic or intermolecular bonds.

Classical Electromagnetic Theory

In classical electromagnetic theory, atoms and molecules are considered to contain electrical charges (*i.e.* electrons, ions) which are regarded as oscillating about positions of equilibrium, each with its appropriate natural frequency, o, When placed in a radiation field of frequency , each oscillator in the atom or molecule is set into forced vibration with the same frequency as that of the radiation.

If << o or >> o , the amplitude of the forced vibration is small, but as approaches the amplitude of the forced vibration increases rapidly. To account for the absorption of energy from the radiation field, it is necessary to assume that the oscillator in the atom or molecule must overcome some frictional force proportional to its velocity during its forced motion.

For small amplitudes of forced oscillation, when is very different from o , the frictional force, and therefore the absorption of energy, is negligible. Near resonance (i.e is approx. equal to o), the amplitude of oscillation becomes large, with a correspondingly large absorption of energy to overcome the frictional force. Therefore, the radiation of frequencies near the natural frequency of the oscillator corresponds to an absorption band.

Acoustic and Optical Modes

Now, let us examine the allowed vibration modes in a lattice. The most rudimentary lattice we can consider is a linear chain of identical atoms. If we displace the nth atom from its equilibrium position by a small amount πn parallel to the chain, the neighbouring atoms ((n-1)th and (n+1)th) will also be displaced by,

πn – πn – 1 and πn – πn + 1 respectively. Therefore the nth atom is actually displaced by an amount πn – πn – 1 with respect to the (n-1th) atom, and by πn – πn + 1 with respect to the (π + 1)th atom.

According to the Lorentz model, these atoms can be considered to be connected by springs. Therefore the n th atom will experience a restoring force F equivalent to that of Hooke's Law. Because the n th atom is connected to two springs (one to each neighbouring atom) there will be two contributions to this force, which is given by:

$$F = -\eta\ [(\pi n - \pi n - 1) - (\pi n + 1 - \pi n)]$$

where η is the spring constant. Applying Newton's second law we also get:

$$F = m\ .\ d2\pi n/dt2$$

where m is the mass of the oscillator. Thus,

$$d2\pi n/dt2 = (\eta/m)\ [(\pi n - \pi n - 1) - (\pi n + 1 - \pi n)]$$

Solutions to equation have the form:

$$\pi n = \pi o\ \ ei(kna - \omega t)$$

where is the angular frequency of the wave, k is the wavenumber and πo is the amplitude of oscillation.

If we now substitute equation we get:

$$\omega 2 = 2(\eta/m)\ [1\text{-}\cos(ka)]$$

$$\omega 2 = 4(\eta/m)\ \sin 2(ka/2)$$

$$\omega = \pm 2\ (\eta/m)\ \tfrac{1}{2}\sin(ka/2)$$

This equation gives the allowed vibration modes in this simple lattice. From equationit is obvious that $\omega > 2(\eta/m)^{1/2}$ is not possible. Frequencies above $\omega = 2(\eta/m)^{1/2}$ cannot propagate through the lattice. In this model the displacement of the oscillator is along the chain, therefore the wave motion is longitudinal. This is the longitudinal acoustic mode.

Now consider a chain in which there are two different atoms of type X and Y, with masses mX and mY respectively (mX > mY), placed alternately along the chain. Again we displace the nth atom, of type X, by a small amount πn. The nearest neighbouring atoms, both of type Y, are displaced by πn-1 and πn+1 as before; the next nearest atoms, both of type X, are displaced by πn-2 and πn+2 and so on. We have two different atom types in the chain, so we require two equations for the restoring force: one for the nth atom of type X and one for the (n+1)th atom of type Y.

Hence we get:

$$FX = -\eta\ X[(\pi n - \pi n\text{-}1) - (\pi n + 1 - \pi n)]$$

$$FY = -\eta Y[(\pi n - \pi n{+}1) - (\pi n{+}1 - \pi n{+}2)]$$

For simplicity, let $\eta X = \eta Y = \eta$. Again we can use Newton's second law and get:

$$FX = mX\ .\ d2\pi n/dt2$$

$$FY = mY\ .\ d2\pi n{+}1/dt2$$

And again, analogous to the monatomic chain we get solutions of the form:

$$\pi n = \pi X\ \ ei(kna - \omega t)$$

$$\pi n = \pi Y\ \ ei(k(n + 1)a - \omega t)$$

From which we get:

$$\omega 2 = \eta[(1/mX) + (1/mY)] \pm \eta[\{(1/mX) + (1/mY)\}2\ - \{4\sin 2(ka)\}/mXmY]$$

If we take the negative sign in equation, we get the acoustic mode already encountered, with the maximum allowed frequency at $= 2(\eta/mX)^{1/2}$ again. However, if we take the positive sign in equation, we get the optical mode, so called because this mode is easily excited by radiation of the appropriate frequency in ionic solids. In this case we find that there is a range of permitted frequencies with both a maximum and a minimum which correspond to the longitudinal and transverse optical (LO and TO) modes.

The minimum frequency (TO) is given by:

$$\omega TO = 2(\eta/mY)^{1/2}$$

The maximum frequency (LO) is given by:

$$\omega LO = 2[\eta/\{(1/mX)+(1/mY)\}]^{1/2}$$

The transverse and longitudinal optical modes represent the onset and cutoff of an absorption band. It is generally thought that absorptivity measurements only involve the TO component, however, in reality the LO component can play an important role (Berreman 1963).

The contribution from the LO mode depends on the constituents of the molecule/solid under consideration as well as the crystal structure, grain size/sample thickness etc. The conditions under which the spectrum is taken and the nature of the spectrometer also affect the relative contributions from the TO and LO modes. It is therefore necessary to understand the contributions from all these factors in order to interpret spectra correctly.

Electrical Properties of Matter

At the beginning of this section the mechanism by which electromagnetic radiation loses energy to material media was outlined. We have just concluded a simple discussion of the permitted frequencies in a material medium. Now we need to combine these two: we have a system in which atoms or molecules are connected by "springs" and that are forced to oscillate in an electric field. The motion is affected by friction in the guise of a damping force. If the oscillator has mass m and charge q, we can expect an equation of motion of the form:

$$m (d2\pi/dt2) + b (d \pi/dt) + \eta\pi = qE$$

where π is the displacement of the oscillator, b is the damping force, η is the spring constant and E electric field of the electromagnetic radiation incident on the material. The electric field is assumed to be time harmonic with frequency ω. From equation we can see that η is approximately equal to m ω o2 (where ωo is the natural frequency of the medium); the damping force can be written as b = γm, where γ is the damping coefficient.

Thus, equation becomes:

$$d2\pi/dt2 + \gamma(d \pi/dt) + \omega o2\pi = qE/m$$

The solution of equation is:

$$\pi = (-qE/m) . [1 / \{(\omega o2 - \omega 2) - i\gamma \omega\}]$$

which gives us an induced dipole moment, p, of:

$$p = (-q2E/m).[1/\{(\omega o2 - \omega 2) - i\gamma \omega\}]$$

If we assume that the displacement π is small we can also assume that there is a linear relationship between the induced dipole moment and the electric field, which gives:

$$p = a(\omega)E$$

where a(ω) is the frequency dependent polarizability of the atom or molecule. Combining equations:

$$a = (-q2/m) . [1/\{(\omega o2 - \omega 2) - i\gamma\ \omega\}]$$

Because of the inclusion of the damping term, the polarizability is a complex quantity. If there are N oscillators per unit volume, the macroscopic polarizability, P, is given by:

$$P = N < p > = N\ a\ E$$

From Maxwell's equations we can get a complex displacement, D, given by:

$$D = E = E + 4P$$

where is the complex dielectric function. Substituting for P from equation we get:

$$E = E + 4N\ a\ E$$
$$= 1 + 4N\ a$$

now substituting from equation we get:

$$= 1 + (4N\ q2)/m.[1/\{(\omega o2 - \omega 2\ ) - i\gamma\ \omega\}]$$

The dielectric function is complex and can, therefore, be written as:

$$= \in' + \in''$$

From equation we get:

$$\in' = 1 + (4N\ q2)/m.[(\omega o2 - \omega 2\ )/\{(\omega o2 - \omega 2)2 + \gamma 2\ \omega 2\}]$$
$$\in'' = 1 + (4N\ q2)/m.[(\gamma\omega)/\{(\omega o2 - \omega 2\ )2 + \gamma 2\ \omega 2\}]$$

We have expressions for the real and imaginary parts of the complex dielectric function (and). What do these expressions tell us about the response of a material to electromagnetic radiation? Let us consider the frequency dependence of and, The frequency dependences given by equations which shows that, for most of the spectrum, increases with increasing frequency, This is known as normal dispersion.

However, for a small section close to the resonance frequency of the oscillator, o, the value of decreases with increasing, This is known as anomalous dispersion. The width of the region of anomalous dispersion is determined by the strength of the damping coefficient,, For small values of , the region of anomalous dispersion is narrow.

However, for large values of , the anomalous dispersion covers a broader range of frequencies, and hence the region over which absorption by the medium extends is broader. Inspection of equations also shows us that, for small values of , the value of remains small and the peak of is very close to the resonance frequency of the medium, o. Likewise, has its anomalous region centred at about o.

However, for large values of , the situation is quite different. With increasing , the height of the peak of becomes much larger and the peak position moves away from the resonance frequency, as does the position of the anomalous region in, Since is the measure of the damping within a medium, *i.e.* the mechanism by which the medium absorbs energy from electromagnetic radiation, large values of correspond to highly absorbing media.

Similarly, low values of correspond to relatively transparent media. Therefore, for relatively non-absorbing materials, a region of anomalous dispersion, and therefore an absorption feature, coincides approximately with the resonance frequency of the oscillators.

Furthermore, for highly opaque, absorbing materials, the anomalous dispersion and the absorption feature will be shifted away from the resonance frequency of the oscillators.

As can easily be seen from equation the physical property of a solid that determines its dielectric function is its electric polarizability, which, in turn, consists of contributions from the individual dipole moments of the oscillators within the solid. It is therefore useful to consider what actually constitutes these dipole moments.

There are three possible sources of individual dipole moments in solids to examine:

1. The dipole moments due to the orientation of polar molecules in the solid;
2. Ionic polarisation due to the relative movement of the ions in the solid;
3. Electronic polarisation due to the displacement of atomic electrons relative to the nuclei. Depending on the nature of the solid, the first two considerations may not be applicable (*i.e.* the solid may not contain polar molecules or ions), however all solids contain electrons and nuclei and therefore electronic polarisation is found in all solids. The contributions to the dielectric function from these three types of "oscillators" lead to features in three separate regions of the spectrum. The dipole moments of polar molecules are expected to give rise to features in the ~1011Hz (~3mm) region of the spectrum; likewise the ionic polarisation affects the spectrum in the ~1013Hz (30μm) region (the infrared); and the electronic polarisation will make contributions at ~1015 (0.3μm; UV). Therefore we expect some solids to exhibit spectral features in all these regions.

REFRACTIVE INDICES OF MATERIAL

There are two sets of optical constants that are closely interrelated.

These are:

1. The real and imaginary parts of the complex refractive index n = n – ik;
2. The real and imaginary parts of the complex dielectric function (relative permittivity) = +,

They are related by:

$\in' = n2 - k2 = Re\ [n2]$

$\in'' = 2nk = Im\ [n2]$

Which of these two sets of constants are most applicable depends on the material under investigation. Reflection and transmission by bulk media are best described using the complex refractive index (n – ik), whereas absorption and scattering by particles which are small compared with the wavelength are best described by the complex dielectric function = $\in' + \in''$, however, they remain interchangeable. The media with which we are concerned have complex refractive indices that depend on wavelength and are complex.

We can write the refractive index, n , of a medium as:

$$n = n - ik$$

To simplify the problem we assume that we have radiation incident on our medium in the form of an idealised, sinusoidal wave.

The incident radiation can be written as:

$$A = Aoe\ i(\omega t - 2x)$$

where A is the amplitude, is the angular frequency and the wave is propagating in the positive x-direction.

Once the radiation enters the medium, the velocity of light becomes c/ n, so that:

$A = Aoe\ i(\omega t - 2n\ x)$

$A = Aoe\ i\omega t - 2i(n - ik)\ x)$

$A = Aoe\ i(\omega t - 2n\ x\ /\ )\ e\text{-}2\ kx$

From this we can see that the intensity of the wave (which is proportional to A2) will decrease exponentially with decay constant of 4k. Therefore the imaginary part of the refractive index, k, is directly responsible for the absorption of the light in the medium.

Scattering

Another consideration in the spectra of small particles is scattering. Consider a small particle placed in a beam of light. The intensity of the elctromagnetic radiation incident on the particle is Io; at a point a large distance r from the particle, the intensity of the scattered light is I. The scattered intensity I must be proportional to the incident intensity Io, and the inverse square of the distance 1/ r 2, hence we can say that:

$$I = [Io\ F(\theta, \phi]/[k\ 2\ r\ 2]$$

where k = 2 is the wavenumber, and $F(\theta, \phi)$ is an orientation dependent, dimensionless function of the direction but not distance from the particle. The total energy scattered by the particle in all directions is defined as being equal to the energy of the incident light falling on the cross-sectional area Csca, where Csca is the scattering cross section. If the function $F(\theta, \phi)$ is divided by k2Csca the phase function is obtained, which has no dimensions and whose integral over all directions is 1.

Therefore:

$$Csca = (1/k2).\S\ F(\theta, \phi)\ d\omega$$

where $d\omega = \sin\theta\ d\theta d\phi$ is the element of solid angle and the integral is taken over all directions. In a similar way to the scattering cross-section case, the energy absorbed by the particle can be put equal to the energy incident on the area Cabs (the absorption cross-section). Likewise the total energy depleted from the original beam can be put equal to the incident energy on the area Cext, the extinction cross-section. The law of conservation of energy requires that:

$$Cext = Cabs + Csca$$

The efficiency of a particle to absorb, scatter and extinguish are given by the Q factors, where:

Cext = Qext x area

Cabs = Qabs x area

Csca = Qsca x area

Therefore:

$$Qext = Qabs + Qsca.$$

The Q-factors can be calculated from the complex refractive index (or complex dielectric function) using Mie Theory

Mie Theory

Mie theory describes the way in which spherical, homogeneous particles interact with electromagnetic radiation. Certain assumptions are made about these particles in order to simplify the situation.

These assumptions are:

- Only the interactions of a single particle with light of arbitrary wavelength are considered.
- The optical properties of the particle are completely described by frequency-dependent optical constants (*i.e.* the complex refractive index or the complex dielectric function).

- Scattering is elastic, *i.e.* the frequency of the scattered light is identical to that of the incident light.
- Only single-scattering occurs. That is, each individual particle is acted on by an external field (due to the incident radiation) in isolation from the other particles. The total scattered field is merely the sum of the fields scattered by each particle (*i.e.* the particles do not affect each other).
- For a collection of particles, the number of particles is large and their separations are random, so that the waves scattered by the individual particles have no systematic phase relation.
- The medium in which the particles are embedded is considered to be:
  a. Linear,
  b. Homgeneous;
  c. Isotropic.

The theory used to find the transmission spectra of small particles from the optical constants is known as Mie Theory. This calculates the Q-factors, which are the efficiency factors for extinction, scattering and absorption. There is no simple explicit formula for the Q-factors, however if 2a, (where a is the radius of a spherical particle and is the wavelength of the incident light), then for a spherical particle:

Qsca = (8/3)(2a/)4 Re[{(n2 – 1)/(n2 + 2)}2]

Qabs = 8a/ Im[(n2 – 1)/(n2 + 2)]

where n = n – ik.

Given the close relation between the complex refractive index and the complex dielectric function, we can also write these equations in terms of the dielectric function:

Qsca = (8/3)(2a/)4 Re[{(epsilon - 1)/(epsilon + 2)}2]

Qabs = 8a/ Im[(epsilon – 1)/(epsilon + 2)]

where epsilon = – i. There are computer programmes that use Mie Theory to evaluate the Q-factors and thus produce the relevant spectra.

These programmes require the input of:

- A dimensionless length 2 8a;
- The real and imaginary parts of the refractive index of the materials n = n – ik, or the real and imaginary parts of the dielectric function of the material epsilon = ∈¢ –∈^{2}.

Bibliography

Aaron Ihde: *The Development of Modern Chemistry*, New York: Dover, 1984.

Anil Kumar: *A Comprehensive Text Book of Applied Chemistry*, Abhishek Pub, Delhi, 2009.

Antwerpen, F.J. : *The Origins of Chemical Engineering, History of Chemical Engineering*, Washington D.C.: ACS, 1980.

Arun Sarkar: *Advanced Organic Chemistry : Reactions and Mechanisms*, Swastik Publications, Delhi, 2011.

Arvind N. Shukla: *Inorganic Biochemistry*, Discovery, Delhi, 2010.

Chambers, K. L.: *Biochemical Coevolution: Eugene*, Oregon State University Press, Biology Colloquium, 1970.

David M.: *Basic Principles and Calculations in Chemical Engineering*, Prentice-Hall, 1996.

Eduard Farber: *Great Chemists*, New York, Interscience Publishers, 1961.

Geis, I.: *Chemistry, Matter, and the Universe*, Menlo Park, Ca., W.A. Benjamin, 1976.

Goel, A K : *Basic Concept of Animal Chemistry*, Pearl Books, Delhi, 2008.

Green, D.W. : *Perry's Chemical Engineers' Handbook*, Eighth Edition, McGraw-Hill, 2007.

Gunjan Goel : *Applied Dairy and Food Microbiology*, Agrotech, 2005.

Hayes, Williams : *American Chemical Industry: Background & Beginnings*, New York: D. Van Nostrand Company: 1954.

Himmelbau, David M.: *Basic Principles and Calculations in Chemical Engineering*, Prentice-Hall, 1996.

Howard, A K : *Aquatic Environmental Chemistry*, Oxford University Press, Delhi, 1998.

Ihde, Aaron John: *The Development of Modern Chemistry*. Courier Dover Publications, 1984.

Jack Simons: *An Introduction to Theoretical Chemistry*, Cambridge Univ Press, Delhi, 1998.

Jai Shankar Ojha: *Aquaculture and Nutrition and Biochemistry*, Agrotech, Delhi, 2006.

John Hays : *The Chemical Engineer, The Profession of Engineering*, New York: John Wiley & Sons, Inc., 1929.

John Hudson: *The History of Chemistry*, New York, Chapman & Hall, 1992.

Johnson-Green, Perry : *Introduction to Food Biotechnology*, Boca Raton, FL: CRC Press, 2002.

Leena Parihar: *Dairy Microbiology*, Agrobios, Delhi, 2008.

Lehninger, A. L.: *Biochemistry*, New York, Worth and Co., 1970.

Mahboob, Syed : *Handbook of Fruit and Vegetable Products*, Agrobios, Delhi, 2008.

Mahinder Singh: *A Textbook of Biochemistry*, Dominant Pub, Delhi, 2011.

Mukherjee, S. : *A Textbook of Objective Organic Chemistry*, Wisdom Press, Delhi, 2011.

Naren Kumar Dutta: *Fundamentals of Biochemistry : A Practical Approach*, Kanishka, Delhi, 2005.

Nicholas A. Peppas: *One Hundred Years of Chemical Engineering*, Boston, Kluwer Academic Publishers, 1989.

Nitin Suri: *Molecular Biology and Biochemistry*, Oxford Book Company, Delhi, 2010.

Patricia Trueman: *Nutritional Biochemistry*, MJP Pub, Delhi, 2007.

Paul B. : *Food Biotechnology in Ethical Perspective*, Aspen, CO: Aspen Publishers, 1997.

Peter J. T. Morris: *The Development of Plastics*, Cambridge, Royal Society of Chemistry, 1994.

Pooja Bhagwan: *A Handbook of Inorganic Chemistry : Nuclear, Atomic, Aqueous, Alkali and Alkaline*, ISPA, Delhi, 2005.

Prasad, S.K. : *Biochemistry of Proteins*, Discovery, Delhi, 2010.

Raghunath Narvekar: *Molecular Biochemistry : Principles and Practices*, Adhyayan Pub, Delhi, 2008.

Ram Naresh Mahaling: *Basics of Biochemistry*, Anmol, Delhi, 2008.

Ranpal Singh: *Advanced Inorganic Chemistry*, RBSA Pub, Delhi, 2006.

Robin Whyman: *Applied Organometallic Chemistry and Catalysis*, Oxford University Press, Delhi, 2005.

Roy Porter: *The Biographical Dictionary of Scientists*, New York, Oxford University Press, 1994.

Ruchi Gautam: *A Textbook of Microbiology*, Arise, Delhi, 2007.

Samuel Delvin: *Science and Technology : Biochemistry*, Sarup, Delhi, 2007.

Samuel Delvin: *Science and Technology : Biochemistry*, Sarup, Delhi, 2007.

Saxena, P B : *Organic Biochemistry*, Discovery, Delhi, 2007.

Shubhrata R. Mishra: *Plant Biochemistry*, Discovery, Delhi, 2010.

Singh, G : *Chemistry of Amino-Acids and Proteins*, Discovery, Delhi, 2007.

Singh, Mahinder: *A Textbook of Biochemistry*, Dominant Pub, Delhi, 2011.

Siradeshmukh, D B and K G Subhadra: *Experiments in Solid State*, South Asian Pub, Delhi, 2008.

Smriti Ganguly: *Biochemistry of Biomolecules*, Pearl Books, Delhi, 2007.

William F. Furter: *A Century of Chemical Engineering*, New York, Plenum Press, 1982.

Wyndham D. Miles: *American Chemists and Chemical Engineers*, Washington, D.C., American Chemical Society, 1976.

Yadav, R.K. : *A Textbook of Objective Inorganic Chemistry*, Wisdom Press, Delhi, 2011.

Index

❑❑❑